Probl.... ...
Practical Advanced Level
Biology

P.W. Freeland

B.Sc., M.Phil., Dip. Ed. F. I. Biol

Head of Science, Worth School, Sussex

HODDER AND STOUGHTON
LONDON SYDNEY AUCKLAND TORONTO

Also by P. W. Freeland
Problems in Theoretical Advanced Level Biology

Acknowledgements

The photographs reproduced in this book were supplied by Heather Angel (pages 7, 44, 85, 87 and 89), Biophoto Associates (pages 55, 63, 67 and 68, bottom left), K. G. Brocklehurst (pages 66, 68, bottom right, 73, 75 and 81) and E. R. W. Campbell (pages 5 and 53).

British Library Cataloguing in Publication Data

Freeland, P. W.
   Problems in practical advanced level biology.
   1. Biology—Examinations, questions, etc
   I. Title
   574'.076   QH316

   ISBN 0 340 35168 3 (Test)

First printed 1985
Second Impression 1985

Typeset in 10/11 pt Univers (Monophoto) by Macmillan India Ltd, Bangalore

Printed in Great Britain for
Hodder and Stoughton Educational,
a division of Hodder and Stoughton Ltd,
Mill Road, Dunton Green, Sevenoaks, Kent TN13 2YD
at the University Press, Oxford.

Preface

This book contains a collection of practical exercises suitable for students of Advanced level biology. A number of the exercises require observation and description of biological materials; others are investigations intended to occupy a single practical session, while the remainder are experiments yielding results over a longer period of time, from several hours to about a week.

Each exercise, presented in the form of a problem, contains a brief statement of background information and lists materials and names the suppliers of unfamiliar items. No attempt has been made to produce a comprehensive collection of experiments covering all aspects of the syllabus: rather, this is a collection of largely unfamiliar material, intended to extend the range of practical work available to students. A number of the experiments involve colorimetric and turbidimetric methods, researched by the author in the period 1970–80. Some of the experiments have been developed from proposals originally published in *The School Science Review* and *Journal of Biological Education*. Detailed instructions are provided in all the exercises in the hope that this will enable students to obtain meaningful and rewarding results. The results published in the complete book are those obtained by students, working under the author's supervision: no claims are made as to the accuracy of these results.

Proposals for further research, based on extensions of experimental procedures outlined in the text, are listed at the end of most of the exercises. It is hoped that some of these proposals will provide students with ideas for individual projects, which can provide valuable experience of learning through the experimental approach.

I would like to express my thanks to the Scientific Research in Schools Committee of the Royal Society, who, through generous provision of a grant, enabled me to undertake research into colorimetric and turbidimetric methods of analysis, with particular reference to the application of these techniques in the teaching of biology at Advanced level. More especially, thanks are due to my supervisor, Professor J. F. Sutcliffe, University of Sussex, for his help and encouragement over many years, and to Miss I. W. Preston, Miles Laboratories Ltd, for technical advice on a number of products manufactured by the Ames Corporation.

Additionally, I am most grateful to my laboratory assistant, Mrs M. Moore, for preparing and testing many of the experiments contained in this book, and to my students, both past and present, for their part in obtaining results.

P.W.F.

Advice to Students

Biology is a practical subject and should, whenever possible, be treated as such. Without an experimental approach little, if any, new information about organisms would become available. Whilst very little front-line research can be carried out in schools, one objective of practical work at the school level is to make students familiar with as wide a range of organisms as possible, introducing different techniques that can be used to study the various processes that occur within these organisms. A second objective is to encourage accurate observation and recording, together with skill in the handling of particular pieces of apparatus. A third and perhaps the most important objective is to encourage an experimental approach, which depends on a particular attitude of mind, efficient organisation and manipulative skill.

Observation

Biologists at A level are required to observe, compare and contrast a variety of organisms, organs, tissues and cells. Both with and without the microscope, practice is required in observation, recording what is observed in the form of drawings, tables and continuous prose.

Class experiments

Some of the practical work carried out in schools consists of well-tried model experiments, of predictable outcome. Such work, apart from illustrating the applications of particular pieces of apparatus or techniques, provides an opportunity for writing a full report of investigations, including the methods employed and the results obtained. It is important that whenever experimental work is described, a particular formal presentation of material should be followed.

(i) Title.
(ii) Objectives: a brief statement of aims.
(iii) Materials: a list of requirements.
(iv) Methods: a logical, ordered statement of procedures.
(v) Results: the outcome of experimental procedures, presented in the form of tables, graphs etc.
(vi) Discussion: including difficulties in the handling of apparatus, limitations of the technique, conclusions and proposals for further research.

Projects

The majority of the Examining Boards will accept a project as an optional part of the A level examination, for which a candidate can gain additional credit. It is a worthwhile venture, as no one who completes a project has anything to lose. Quite apart from gaining experience in methods of scientific investigation, a project can provide a useful talking point at interview, or help a mediocre examinee to gain a few extra marks.

If you do decide to tackle a project, bear the following points in mind.

(i) Begin your project during the first year of study, aiming to have it completed before your last term.
(ii) Don't be too ambitious or unrealistic in your choice of subject. There are obvious limitations, both academic and economic, to the type of problem that can be investigated at school or in the home.
(iii) Attempt to choose a subject in which you have a genuine interest. Remember that credit is awarded for originality, clarity of presentation and for the scientific (or economic) merits of your work. The inclusion of photographs can be helpful, particularly if they provide visual evidence in favour of your conclusions.
(iv) If you are unable to come up with any ideas of your own, you may gain some ideas by reading through the proposals at the end of each exercise.

Using this book

Sections of this book listed under Preparation and Materials are intended for your teacher or supervisor. In cases where unfamiliar chemicals are required, reference numbers[1,2] etc. refer to suppliers listed on p. 147.

Contents

1 Identification and Estimation of Some Chemical Compounds

A colorimetric method for the estimation of glucose (or reducing sugars) in solution

Glucose, a reducing agent, will reduce an acidified purple–pink solution of potassium manganate(VII) (potassium permanganate) to a colourless solution of manganese(II) ions. The partial ionic equation is

$$MnO_4^- + 8H^+ + 5e^- \rightarrow Mn^{2+} + 4H_2O$$

Purple–pink   *Colourless in*
in solution   *solution*

The time taken for loss of colour from a standardised solution of permanganate is directly related to the concentration of glucose present in solution. After plotting a standard curve for different concentrations of glucose, all of which are known, you are required to use your graph to estimate the concentrations of glucose in five solutions, labelled A to E, provided by your supervisor.

Preparation

The standardised glucose solution contains 10 g glucose dissolved in 100 cm³ water. If the exercise is to be presented as a problem, the supervisor must provide five additional glucose solutions, labelled A to E, each containing a known mass of glucose dissolved in 100 cm³ distilled water.

Materials

- Fifteen flat-bottomed tubes, each approximately 2 ×8 cm
- Three 10 cm³ plastic syringes
- 150 cm³ glucose solution
- 60 cm³ 1 M sulphuric acid
- 30 cm³ 0.01 M potassium manganate(VII) solution
- 150 cm³ distilled water
- 10 cm³ of each glucose solution, A to E
- Stop-watch

Method

1. Using the plastic syringe, introduce 10 cm³ glucose solution and 5 cm³ sulphuric acid into a flat-bottomed tube. Draw 2 cm³ potassium manganate (VII) solution into a syringe, and press the stop-watch at the moment this addition is made to the contents of the tube. Record the time, in seconds, for complete decolourisation of the potassium manganate (VII).
2. Using the syringes, make dilutions of the glucose solution provided to obtain solutions containing, respectively, 9, 8, 7, 6, 5, 4, 3, 2, and 1 g glucose per 100 cm³ distilled water.
3. Repeat procedure 1 with each solution of glucose, recording the time, in seconds, taken for complete decolourisation of the potassium manganate (VII). Record all your results in the form of a table and plot your results as a graph.
4. Use the graph to estimate concentrations of glucose in each of the solutions labelled A to E. Record your results.
5. Comment on your results.

Topics to investigate

1. Changes in the glucose content of ripening fruits, such as grapes.
2. Estimation of reducing sugars in fruit juices.
3. Estimation of reducing sugars in white wines.
4. Rates of glucose assimilation by yeast or bacteria.
5. Investigations into the activity of invertase.

1.2 Chemical methods for estimating the pH, hardness, salinity and oxygen content of water

Differences in the pH, hardness, salinity and oxygen content of water can exert a marked influence on the variety and numbers of living organisms, both animals and plants, that are able to colonise the water. This exercise introduces methods for measuring some of the more variable chemical properties of water, and draws attention to the chief differences between fresh water, drawn from a pond or river, and sea water.

Preparation

Each student requires at least 300 cm³ pond water and 300 cm³ sea water for this exercise. Twenty-four hours before the exercise is presented, each participant should pour 50 cm³ pond water into each of two 100 cm³ beakers and 50 cm³ sea water into each of two beakers. The beakers of water should stand on the bench surface, at room temperature, until they are required for analysis.

The manganese(II) sulphate solution is prepared by dissolving 25 g manganese sulphate in 100 cm³ water. Alkaline potassium iodide reagent contains 50 g sodium hydroxide and 20 g potassium iodide dissolved in 100 cm³ water.

Materials

- Two 100 cm³ beakers, each containing 50 cm³ pond water
- Two 100 cm³ beakers, each containing 50 cm³ sea water
- 250 cm³ pond water
- 250 cm³ sea water
- 5 cm³ manganese (II) sulphate solution
- 5 cm³ alkaline potassium iodide solution
- 5 cm³ conc. sulphuric acid
- 50 cm³ 0.001M sodium thiosulphate solution
- 20 cm³ 0.1M hydrochloric acid
- Soluble starch solution (indicator)
- Two 100 cm³ beakers
- Two flat-bottomed tubes, approximately 2 × 8 cm
- Two 1 cm³ plastic syringes
- Pestle and mortar
- Quantab chloride titrators no. 1175 and 1177[4]
- Water-hardness tablets[1]
- Whatman or Merckoquant narrow-range pH papers, pH 0–8.5, or a pH meter
- Bunsen burner, tripod and gauze

Method

1 Introduce 100 cm³ pond water into a beaker. Use the narrow-range pH papers to determine the pH of the water. Record your result. Add 10 cm³ 0.1 M hydrochloric acid, in units of 1 cm³, to the water and record the pH of the water after each addition. Repeat this procedure with the sea water. Present all your results in the form of a table. Plot your results as a graph and comment briefly on them.

2 Introduce 100 cm³ pond water into a beaker and add crushed water-hardness tablets until there is a colour change in the mixture from reddish-purple to blue-grey. Record the number of tablets added. Using 100 cm³ water,

$$\text{hardness (p.p.m.)} = (\text{no. of tablets} \times 20) - 10$$

Introduce 10 cm³ sea water into a beaker and add 90 cm³ distilled water. Add crushed water-hardness tablets to this mixture until there is a colour change. Using 10 cm³ water,

$$\text{hardness (p.p.m.)} = (\text{no. of tablets} \times 200) - 10$$

What is the cause of hardness in water?
What is the chief biological significance of the results obtained for plants and animals living in fresh water and sea water?

3 Introduce 2 cm³ pond water into a flat-bottomed tube. Stand a Quantab chloride titrator no. 1175 in the water and read off the salinity after the yellow band at the top of the scale has changed colour from yellow to blue. Record your result, reading it off from the table supplied by the manufacturer. Repeat the procedure with sea water, using Quantab chloride titrator no. 1177. Record your results.

Of what importance is the salinity of water to plants and animals that live in an aquatic habitat?

4 Take the two beakers containing 50 cm³ pond water. Transfer one of the beakers to the tripod and apply heat until the water boils. Rapidly cool the water. Add 0.5 cm³ manganese(ii) sulphate solution, followed by 0.5 cm³ alkaline potassium iodide reagent and mix thoroughly. Allow the mixture to stand for eight minutes, then add 0.5 cm³ conc. sulphuric acid. After mixing, titrate against standardised 0.001M sodium thiosulphate solution, using soluble starch solution as indicator.

  Approximate percentage of oxygen
  $= 12 \times 0.001 \times$ volume of sodium thiosulphate (cm³)

Repeat the procedure with unboiled pond water. Record your results and comment briefly on them.

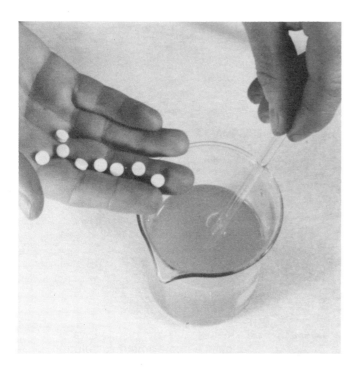

Determining the hardness of pond water. Water-hardness tablets are added and crushed until there is a colour change in the mixture.

5 Repeat procedure 4 using sea water. Record your results, and comment on any differences that exist between the pond water and sea water.

Topics to investigate

1 Changes in the salinity of (i) river water and (ii) rock pools during the ebb and flow of the tides.
2 Effects of increasing salinity on the oxygen content of water.
3 Relationship between the hardness of pond water and the relative density of a named mollusc in different ponds.
4 The pH of pond water as a factor determining the distribution of a named plant or animal.
5 Effects of the hardness of water on rates of photosynthesis in a named aquatic plant.

Nitrogenous compounds such as ammonium (NH_4^+), nitrate (NO_3^-) and nitrite (NO_2^-) ions, urea, amino acids and proteins may occur in a fertile loam soil. Interconversions between some of these compounds, as in the nitrogen cycle, can be demonstrated by using reagent sticks for the semi-quantitative estimations of ions in the soil solution. The nature of these interconversions depends on the presence or absence of oxygen, and may be summarised as proceeding in a particular direction, depending on whether the soil provides conditions favouring the oxidation or reduction of nitrogenous compounds.

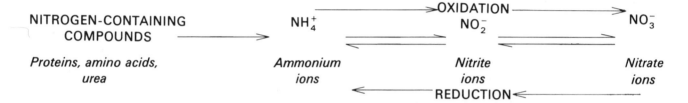

This exercise illustrates the nature of the end products and presents a method for the measurement of rates of end-product formation when different nitrogenous compounds, maintained under different conditions, are incubated with a loam soil.

Preparation

The solution of sodium nitrate contains 1 g solute dissolved in 100 cm³ water, and the solution of urea contains 2 g solute/100 cm³ water.

Materials

- Seven 250 cm³ beakers
- 350 g loam soil
- 1 g sodium nitrite
- 150 cm³ sodium nitrate solution
- 150 cm³ urea solution
- 5 cm³ dilute (bench) ammonia solution
- 300 cm³ distilled water
- Fresh egg
- Merckoquant nitrite-sensitive reagent sticks[1]
- Merckoquant nitrate-sensitive reagent sticks[1]
- Merckoquant ammonium-sensitive reagent sticks[1]
- Top-pan balance
- Water-bath maintained at 30 °C.

Method

1 Set up seven beakers as indicated:

Beaker 1   50 g loam + 150 cm³ distilled water
Beaker 2   50g loam + 1 cm³ ammonia solution
Beaker 3   50g loam + 150 cm³ distilled
                 water + 1 cm³ ammonia solution
Beaker 4   50 g loam + 1g sodium nitrite
Beaker 5   50 g loam + 150 cm³ sodium nitrate
                 solution
Beaker 6   50 g loam + 5 g egg white
Beaker 7   50 g loam + 150 cm³ urea solution

Stir the contents of each beaker to ensure that mixing has taken place. Immediately after setting up the beakers, make a semi-quantitative estimation of concentrations of NO_2^- NO_3^- and NH_4^+ ions in each mixture. From beakers 2, 4 and 6, withdraw 5 g soil for each test. Add 15 cm³ distilled water to each sample and then apply the semi-quantitative tests. Multiply each figure obtained by a factor of 4, to give the concentrations of ions in the soil sample.

2 Transfer all the beakers to a water-bath maintained at 30 °C. Ensure that the soil samples in beakers 2, 4 and 6 remain in a moist, crumbly condition and do not dry out. At intervals of 6 hours, over a period of 24 hours, measure and record the concentrations of NO_2^-, NO_3^- and NH_4^+ ions in each mixture. Present all your results in the form of a table.

Root nodules on the root of a runner bean.

3 Plot a graph to show the nature of the end products, and the rate of end-product formation, in beakers 4 and 5.
4 Plot a graph to show the nature of the end products, and the rate of end-product formation, in beaker 7.
5 Write a brief account of the processes taking place in each beaker and the significance of these processes in the cycling of nitrogen.

1 Short-term and long-term changes in the form and concentration of nitrogen-containing compounds in a soil, following application of artificial fertiliser.
2 Losses of NO_2^-, NO_3^- and NH_4^+ ions from a soil as a result of leaching.
3 Rates of release of NO_3^- and NH_4^+ ions from plant material during composting.
4 The effect of a species of pot-grown legume on the nitrogen status of the surrounding soil.
5 Effects of pH and waterlogging on the rate of denitrification of a soil.

1.4 | A systematic identification of carbohydrates

Reagents are provided for the identification of carbohydrates. The object of this exercise is to devise a scheme for the systematic identification of carbohydrates and to use it to identify three carbohydrates, labelled A to C, provided by your supervisor.

Preparation

The supervisor should provide each student with 1 g of each of three different carbohydrates, labelled A, B and C. Each carbohydrate should be ground to a fine powder in order to camouflage its identity.

Reagents required for this exercise have the following composition:

Barfoed's reagent
6.6 g copper(II) acetate
1 cm³ glacial acetic acid
100 cm³ water

Benedict's reagent
7.3 g sodium citrate
10.0 g anhydrous sodium carbonate in 80 cm³ water
Add to
1.7 g copper(II) sulphate in 20 cm³ water

Bial's reagent
0.3 g orcinol, dissolved in 20 cm³ conc., hydrochloric acid
1 cm³, 0.25 g ferric chloride dissolved in 100 cm³ water
80 cm³ water

Schultze's reagent
10 g zinc, dissolved in 30 cm³ 2 M hydrochloric acid
(*Heat to reduce total volume to 15–20 cm³*)
1.2 g potassium iodide
0.015 g iodine crystals

Seliwanoff's reagent
0.5 g resorcinol, dissolved in 35 cm³ conc. hydrochloric acid
65 cm³ water

Materials

- Three unidentified carbohydrates, labelled A, B and C
- Barfoed's reagent
- Benedict's reagent
- Bial's reagent
- Schultze's reagent
- Seliwanoff's reagent
- Clinistix reagent strips[3, 4]
- Iodine in potassium iodide solution
- 1 M hydrochloric acid
- Boiling tubes
- Bunsen burner
- Glass marking pen

Method

Some of the more common carbohydrates, including A, B and C, are listed in Table 1, while some of the reagents currently used to identify carbohydrates are listed in Table 2.

Table 1 *Some common carbohydrates. (* denotes reducing sugars.)*

Type of carbohydrate	Examples	Solubility in water
Monosaccharides	Glucose*	Soluble
	Fructose*	Soluble
	Xylose	Soluble
Disaccharides	Sucrose	Soluble
	Maltose*	Soluble
Oligosaccharides	Starch	Insoluble
	Glycogen	Insoluble
	Cellulose	Insoluble

Table 2 *Reagents used to test for carbohydrates.*

Reagent	Reaction
Barfoed's	After boiling, rapid formation of a red-brown precipitate with all reducing monosaccharides. The reaction with reducing disaccharides is much slower.
Benedict's	After boiling, a red-brown precipitate with all reducing sugars.
Bial's	After boiling, a green colouration with xylose.
Clinistix	Pink → blue with glucose only.
Iodine	Blue-black with starch, purple-red with glycogen and no reaction with cellulose.
Schultze's	Purple colouration with cellulose.
Seliwanoff's	After boiling, a red colouration with fructose and sucrose.

WARNING: Bial's and Seliwanoff's reagents are acid. Take care when heating carbohydrates with these reagents.

1 Devise a logical, step by step, qualitative scheme for the identification of an unknown carbohydrate. Your scheme should lead, via appropriate tests and positive($+$ve) or negative($-$ve) results to each of the carbohydrates listed.

2 Place each of the samples A, B and C into labelled boiling tubes and add $10 \, cm^3$ water to each sample. Sub-divide each sample into an appropriate number of batches, each of $1-2 \, cm^3$, and use your scheme to identify carbohydrates A, B and C. Record your results.

3 How could you use enzymes and any of the reagents listed to distinguish between finely powdered starch and cellulose?

Topics to investigate

1 A systematic identification of amino acids.

1.5 Estimation of citric acid, ascorbic acid and reducing sugars in the juice of orange, lemon and grapefruit

TIME 2–2½ h

Citric acid, ascorbic acid and reducing sugars are present in juices obtained from citrus fruits. A simple quantitative method for the estimation of each of these compounds is described.

Preparation

The standardised solution of citric acid contains 1 g solute dissolved in 100 cm³ water.

Materials

- Whole orange
- Whole lemon
- Whole grapefruit
- 100 cm³ measuring cylinder
- Access to a top-pan balance, weighing in units of 0.5 g up to 500 g
- Twenty or more dry test-tubes in a rack, or flat-bottomed tubes
- Burette, or 10 cm³ plastic syringe, graduated in 0.5 cm³ units
- Three 10 cm³ syringes
- Nine beakers of 200 cm³, or 250 cm³, capacity
- 1 cm³ plastic syringe
- 25 cm³ citric acid solution
- 200 cm³ 0.1 M sodium hydroxide solution
- 300 cm³ distilled water
- 10 cm³ Universal Indicator[1]
- Universal Indicator colour chart[1]
- Ten 2,6-dichlorophenol indophenol (DCPIP) tablets[1]
- Twenty Clinitest tablets[3,4]
- Clinitest colour chart[3,4]
- Sharp kitchen knife
- Glass rod
- Glass marking pen
- Forceps

Method

1 Weigh each of the citrus fruits on the top-pan balance. Record the mass of each fruit. Cut each fruit in half transversely and squeeze the juice into one of the beakers. Label each beaker. Transfer each of the juices in turn to the measuring cylinder, and measure and record the total volume of juice obtained from each fruit. Return each fruit juice to its appropriate beaker. Assuming that each juice has the same density as water, calculate the percentage of juice in each fruit. Record all your results in the form of a table.

2 Using one of the 10 cm³ plastic syringes, introduce 20 cm³ of the sodium hydroxide solution into each of four beakers. Add 0.5 cm³ (10 drops) of Universal Indicator to each beaker; the colour of the indicator should be a deep purple.

   Fill a second 10 cm³ plastic syringe, or the burette, with citric acid solution. Add the citric acid, drop by drop, to the sodium hydroxide solution in one of the beakers, until the indicator is green, at pH 7.0. Record the volume of citric acid solution added. Repeat this procedure using the juice of orange, lemon and grapefruit. Record the volume of each juice added. If necessary, take a second set of readings in an attempt to obtain a more accurate set of results.

   Assuming citric acid to be the only acid component of the three fruit juices, calculate the percentage of citric acid in each juice.

3 The blue dye 2,6-dichlorophenol indophenol (DCPIP) is reduced by ascorbic acid (vitamin C) to a colourless leucobase. You are supplied with DCPIP tablets, each of which contains a quantity of dye equivalent to 1 mg ascorbic acid. Introduce 10 cm³ distilled water into a beaker and dissolve one DCPIP tablet in it, gently crushing the tablet with the glass rod to ensure that all the dye is in solution. Fill a 10 cm³ plastic syringe with orange juice and add it, drop by drop, to the solution of DCPIP until the colour changes from pink to colourless. Again, if necessary, take a second set of readings to obtain a more accurate result. The volume of juice required to effect this colour change in the indicator contains 1 mg ascorbic acid. Repeat the procedure using lemon juice and grapefruit juice. Calculate (i) the amount of ascorbic acid in each fruit, and (ii) the amount of ascorbic acid per 100 cm³ fruit juice. Present your results in the form of a table.

4 Clinitest tablets provide a convenient method for the semi-quantitative estimation of reducing sugars in fruit juices. As each tablet contains solid sodium hydroxide, the reagent should be handled with forceps only. **Avoid skin contact.**

Introduce 1 cm³ orange juice into a test-tube. Add one Clinitest tablet to the juice. The mixture will boil and its final colour will be blue, green, red-brown or yellow-brown, depending on the amount of reducing sugar present. Compare the final colour with the colour chart and read off the percentage of reducing sugar present (Table 3). If the final colour is yellow-brown, dilute the orange juice with an equal volume of distilled water. Make further dilutions of the juice until a shade of colour, indicating less than 2% of reducing sugar, is obtained. Calculate the percentage of reducing sugar in orange juice. Repeat the procedure using lemon juice and grapefruit juice. Present all your results in the form of a table. Show all the calculations involved in arriving at your answers.

5 State, with reasons, the limitations of the experimental methods employed.

Table 3 *Approximate content of reducing sugars using Clinitest tablets.*

Colour of indicator	% reducing sugar
Blue	0.0
Blue-green	0.25
Green	0.5
Green-brown	0.75
Brown	1.0
Yellow-brown	2.0 or more

Topics to investigate

1 Estimation of citric acid, ascorbic acid and reducing sugars in fruit squashes, drinks etc.
2 Estimation of ascorbic acid and reducing sugars in juice extracted from tomatoes, grapes, apples, potatoes etc.
3 Effects of different methods of extraction, preservation and storage on the ascorbic acid content of orange juice.
4 Breakdown of ascorbic acid by ascorbic acid oxidase. (The enzyme ascorbic acid oxidase, released from fruits when their tissues are crushed and activated by exposure to the atmosphere, causes losses of ascorbic acid from the juice. The enzyme can be inactivated by boiling the juice immediately after extraction. To what extent does this treatment preserve the ascorbic acid, preventing it from being broken down by the enzyme?)
5 Seasonal variations in the ascorbic acid content of oranges purchased from greengrocers.
6 Comparisons of the citric acid, ascorbic acid and reducing sugar content of lemons from different countries of origin.

Distribution of chemical compounds in plant tissues

TIME 1½–2h

Certain chemical compounds that occur in plants are unevenly distributed throughout the organs or tissues in which they occur. As a result of using appropriate stains, applied to sections of plant material, it is possible to visualise and map the distribution of these compounds.

Preparation

Reagents required for this exercise have the following composition:

Stain W Dissolve 2g iodine crystals and 4g potassium iodide in 100 cm³ water.

Stain X Dissolve 2.5 g silver nitrate in 100 cm³ water.

Stain Y Dissolve 0.4 g bromocresol green[1] in 20 cm³ propan-2-ol and 80 cm³ water.
   As an alternative to this stain, dissolve 0.2 g Ponceau S[1] in 100 cm³ water.

Stain Z Dissolve 0.5 g 2,3,5-triphenyltetrazolium chloride[1] in 100 cm³ water.

Materials

- Four slices of cucumber
- Four slices of banana
- Four germinating grains of maize
- Potato, with emergent shoots
- 40 cm³ iodine solution, labelled W
- 40 cm³ silver nitrate solution, labelled X
- 40 cm³ bromocresol green solution, labelled Y
- 40 cm³ 2:3:5 triphenyltetrazolium chloride solution, labelled Z
- 40 cm³ 0.1 M hydrochloric acid
- Ten petri dishes
- Scalpel

Method

1 Fig. 1 shows the appearance of sections through a cucumber (a), banana (b), maize grain (c) and potato (d). Before attempting the exercise, name the tissues illustrated in the four drawings, (a), (b), (c) and (d).

2 Cut the maize grains and potato to obtain sections similar to those illustrated. Pour approximately 20 cm³ of stain W into each of two petri dishes. Transfer one section of each type of plant material to one of the dishes. Allow the dishes to stand for 30 minutes.

3 Repeat procedure 2 with stains X, Y and Z. After 30 minutes transfer material immersed in stain Y to a petri dish containing 0.1 M hydrochloric acid. Leave material in the acid for 5–10 minutes.

4 What compounds are being identified by each of the stains?

5 By means of either a drawing, or a table, attempt to indicate the precise location of compounds within each type of plant material, as demonstrated by the use of stains W, X, Y and Z.

(a) T.S. Cucumber

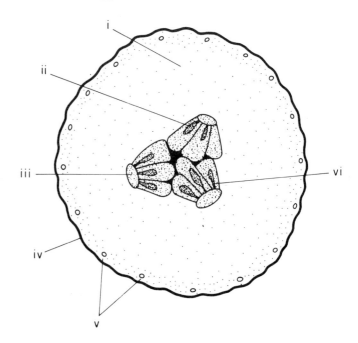

Fig. 1 *Sectioned plant materials*

(b) T.S. Banana

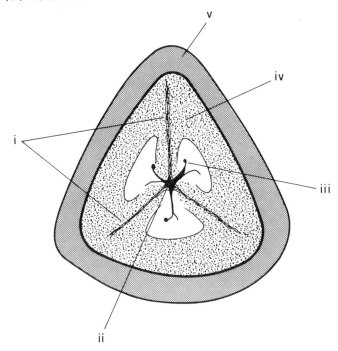

(c) V.S. Germinating Maize Grain

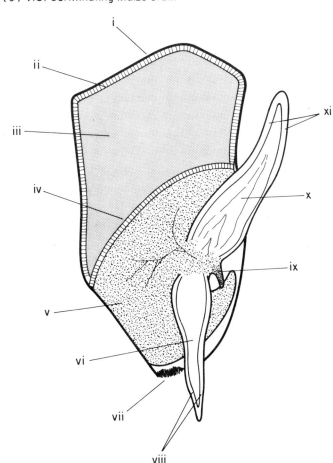

(d) V.S. Potato

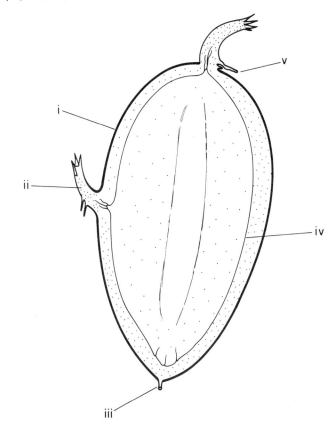

Topics to investigate

1 Changes in the starch-sugar content of bananas during ripening.
2 Mapping the location of stored products in bulbs, corms and tap roots.
3 Changes in the distribution of food reserves during the germination of cereal grains, such as maize.
4 The biochemical basis of staining methods used in (i) plant histology and (ii) animal histology.

1.7 Identification and separation of photosynthetic pigments by thin-layer chromatography

Paper chromatography is widely used to separate and identify photosynthetic pigments. Thin-layer chromatography, however, effects a more complete separation of these pigments and opens up the possibility of a wide range of investigations into the nature of those pigments present in leaves, stems, roots, petals and fruits. This exercise requires an investigation into the variety and distribution of photosynthetic pigments in the leaves and root of carrot, and a report on the separation of these pigments by thin-layer chromatography, using two different solvents.

Preparation

The chromatographic solvents required for this exercise have the following composition:
Solvent A 9 parts petroleum ether: 1 part acetone
Solvent B 5.5 parts cyclohexane: 4.5 parts ethyl acetate
Both solvents should be supplied in stoppered bottles.

Materials

- Young carrot, bearing leaves
- 10 cm³ acetone in a stoppered bottle
- 20 cm³ solvent A
- 20 cm³ solvent B
- Two stoppered tubes
- Four 250 cm³ beakers, or small jars
- Pestle and mortar
- Eight TLC plastic silica gel sheets, each 2.5 × 8 cm[1]
- Two small paint brushes
- Four elastic bands
- Aluminium foil or clingfilm
- Ruler, graduated in millimetres
- Coloured pencils (yellow, orange, yellow-green, blue-green).

Method

1 Remove the leaves from the carrot and discard the petioles. Transfer some of the leafy material to the mortar and grind it to a thick paste. Add approximately 1 cm³ acetone to the paste and continue to grind, then add a further 2 cm³ acetone, grinding as you add the solvent. Pour off the extract into one of the empty tubes. Label the extract.
2 Clean out the mortar and repeat procedure 1 with the root of the carrot. Label the extract.
3 Using one of the paint brushes, apply the leaf extract in a thin, straight line approximately 0.5–1.0 cm from one end of the sheet, to four TLC silica gel sheets. Make several applications, one on top of the other, of the extract to each sheet, allowing 30–60 seconds for each application to dry before the next application is made.
4 Use a second brush to apply the root extract to the four remaining sheets.

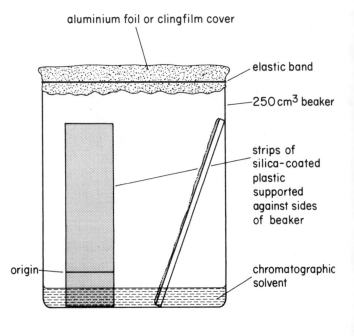

aluminium foil or clingfilm cover

elastic band

250 cm³ beaker

strips of silica-coated plastic supported against sides of beaker

origin

chromatographic solvent

Fig. 2 *Apparatus for separating photosynthetic pigments by thin-layer chromatography*

5 Pour 10 cm³ of solvent A into each of two beakers. Transfer two TLC sheets marked with leaf extract to one of the beakers. Support the sheets against the sides of the beaker at an angle of approximately 45°, with the marked region positioned above the solvent (Fig. 2). Cover the beaker with aluminium foil, or clingfilm, secured in position with an elastic band. Repeat this procedure with two sheets marked with root extract.

6 Follow the procedure outlined in 5 to set up leaf and root extracts in solvent B.

7 Observe the chromatograms as they develop. Remove the chromatograms from the beakers at the point where the solvent front is approximately 0.5 cm from the top edge of each sheet. Make drawings to show the appearance of each chromatogram, using the coloured pencils to show the distribution of the different pigments.

8 What conclusions are you able to draw regarding (i) the effects of the two solvents A and B on the separation of the pigments, and (ii) the distribution of pigments in the leaf and root of the carrot?

9 As many as nine different pigments may separate in solvent B. Carotenes are orange, chlorophylls and chlorophyllides green, and xanthophylls yellow. After drawing the distribution of the pigments from leaves of the carrot, attempt to calculate the R_f value of each pigment.

$$R_f = \frac{\text{distance moved by solute}}{\text{distance moved by solvent front}}$$

After the solvent front has almost reached the top of the TLC sheet, the position of the front is marked. Distances are then measured between the origin and the position of each separate pigment.

Record your results in the form of a table.

10 List four factors that might affect R_f values.

Topics to investigate

1 Synthesis of photosynthetic pigments in roots of carrot exposed to light.
2 Effects of (i) light intensity and (ii) light quality on the production of photosynthetic pigments in seedlings of a named plant.
3 Seasonal changes in the photosynthetic pigments of leaves.
4 Effects of mineral ion deficiencies on the development of photosynthetic pigments in leaves of a named plant.
5 Analysis of pigments from (i) the cuticle of a herbivorous insect and (ii) the plant on which it feeds. (Larvae of the large and small cabbage white butterfly could be used in this investigation.)

1.8 Identification and separation of anthocyanidins by paper chromatography

Purple/red anthocyanidin pigments, like photosynthetic pigments, can be separated and identified by means of paper chromatography. Separation, however, is more difficult to achieve and work is often complicated by the fact that several different pigments may be present in the same organ or tissue.

Preparation

Acidified methanol contains 1 part molar hydrochloric acid: 9 parts methanol. The chromatographic solvent contains 30 parts glacial acetic acid: 3 parts conc. hydrochloric acid: 10 parts water.

If apparatus for paper chromatography is not available, large jars, sealed with aluminium foil or clingfilm, held in place by an elastic band, can be used.

Materials

- Two large jam jars, Kilner jars etc.
- Chromatography paper
- 20–30 cm³ chromatographic solvent
- Flowers of *Pelargonium*, labelled A
- Red rose, labelled B
- Two or three black grapes, labelled C
- Flowers of bluebell (or *Delphinium*), labelled D
- Flowers of *Petunia*, labelled E
- Strawberry, labelled X
- Red-skinned apple, labelled Y
- A slice of red cabbage, labelled Z
- Eight flat-bottomed tubes, approximately 2 × 8 cm
- 20 cm³ acidified methanol
- Two small paint brushes
- Pestle and mortar
- Pencil
- Scissors
- Water-bath maintained at 90 °C.

Method

Each of the five specimens, labelled A to E, contains a different anthocyanidin pigment, as indicated in Table 4. Attempt to determine the R_f values for each of the pigments listed in the table and use this information for the identification of pigments contained in specimens X, Y and Z.

Table 4 *Anthocyanidin pigments.*

Specimen	Anthocyanidin present
Flowers of *Pelargonium* (A)	Pelargonidin
Petals of red rose (B)	Cyanidin
Skins of black grapes (C)	Malvidin
Flowers of bluebell or *Delphinium* (D)	Delphinidin
Flowers of *Petunia* (E)	Petunidin

1 Extract the pigment from each specimen, A to E, in turn, by grinding the material provided in a mortar with 1–2 cm³ acidified methanol. Pour each extract into one of the flat-bottomed tubes provided. Label each tube. Wash out the mortar after preparing each extract.

2 Transfer the tubes to a water-bath maintained at 90 °C, and leave the tubes for 30–40 minutes at this temperature, or until the extract is reduced to a quarter of its original volume.

3 Mark with a pencil a line approximately 1.5 cm from one edge of the chromatography paper. Use a paint brush to convey an extract containing pelargonidin to the paper. Apply the pigment as a small spot along the pencil line. Allow the pigment to dry and then apply a second layer of pigment on top of the first. Thoroughly wash the brush before applying a spot containing cyanidin, spaced approximately 1 cm from the first spot. Repeat this procedure until each pigment has been transferred to the paper. Use the pencil to label each pigment.

4 Prepare extracts from specimens X, Y and Z, spotting these on to chromatography paper and labelling them with a pencil.

5 Carefully pour 20–30 cm³ chromatographic solvent into the jar, taking care not to spill any of the solvent, which contains strong acids. Transfer the chromatography paper, spotted with identified and unidentified pigments, to the jar, ensuring that the solvent is below the level of the pigments. Seal the jar and allow 3–6 hours for development of the chromatogram. Attempt to calculate the R_f value of each pigment.

$$R_f = \frac{\text{distance moved by solute}}{\text{distance moved by solvent front}}$$

Record your results.

6 Calculate R_f values for the pigments contained in the three specimens X, Y and Z. What do you conclude about the identity of these anthocyanidins?

1 Anthocyanidin pigments in flower petals, fruits and vegetables.
2 Effects of different grades of chromatography paper on (i) the rate and (ii) the resolution of separation of anthocyanidin pigments.
3 Anthocyanidin pigments and the classification of plants.
4 Separation of anthocyanidin pigments by thin-layer chromatography.

This chemical analysis of wines involves an estimation of total acidity, percentage ethanol by volume, and an identification, by paper chromatography, of the organic acids present.

Preparation

The chromatographic solvent for this exercise contains 60 parts butan-1-ol: 15 parts glacial acetic acid: 25 parts water.

Materials

- 200 cm³ white wine, labelled A
- 200 cm³ pale sherry, labelled B
- 50 cm³ 0.1 M sodium hydroxide solution
- Phenolphthalein indicator
- Two 100 cm³ conical flasks
- Burette
- 50 cm³ absolute ethanol
- 800 cm³ distilled water
- Two 100 cm³ weighing bottles
- Top-pan balance
- 'Quickfit' distillation apparatus
- Bunsen burner, tripod and gauze
- Two 100 cm³ measuring cylinders
- Two 10 cm³ plastic syringes
- 1 cm³ plastic syringe
- Large Kilner jar and 100 cm³ beaker
- 50 cm³ chromatographic solvent
- 0.5 g bromocresol green in 100 cm³ 0.05 M sodium hydroxide solution
- Chromatography paper
- Ruler, graduated in millimetres
- Plasticine
- Solutions of succinic, lactic, malic, citric and tartaric acids, each containing approximately 1 g acid dissolved in 10 cm³ distilled water
- Scissors, paint brush and pencil

Method

1 Transfer the sodium hydroxide solution to the burette. Introduce 10 cm³ of wine A into a conical flask, add a few drops of phenolphthalein indicator, and titrate against the sodium hydroxide solution until a faint pink colour persists after shaking. Record the total acidity of the wine. (Total acidity is defined as the volume of 0.1 M sodium hydroxide required to neutralise 100 cm³ of wine).

Repeat this procedure with wine B. Record your result.

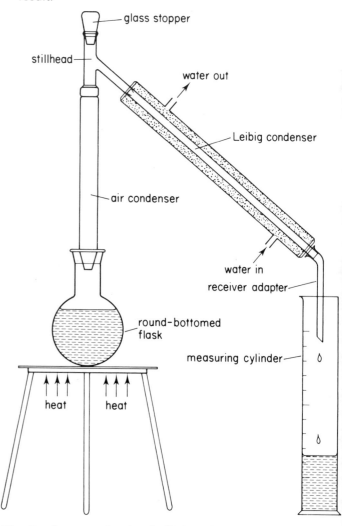

Fig. 3 *Apparatus for the distillation of wine*

2 Assemble apparatus for distillation as illustrated in Fig. 3. Introduce 100 cm³ wine A into the round-bottomed flask and apply heat until 50–60 cm³ distillate has collected in the measuring cylinder. Add distilled water to both the distillate and residue of wine A to restore the total volume of each fraction to 100 cm³. Retain the two fractions and label them. Repeat the procedure with wine B.

3 Prepare ethanol/water mixtures containing, respectively, 5, 10, 15 and 20% ethanol by volume. Transfer 100 cm³ of each mixture to a weighing bottle and record the specific gravity of each mixture. Record your results as a graph. Similarly, prepare solutions of glucose containing, respectively, 5, 10, 15 and 20 g glucose dissolved in 100 cm³ distilled water. Record the specific gravity of each solution and record your results as a graph.

4 Transfer 100 cm³ distillate from wine A to a weighing bottle. Record the specific gravity of the distillate and estimate, from your graph, the volume of ethanol in 100 cm³ of the wine. Transfer 100 cm³ of the residue from wine A to a weighing bottle. Record the specific gravity of the residue and estimate, from your graph, the approximate mass of glucose in 100 cm³ of the wine. Record your results. Repeat the procedure with wine B. Record your results.

5 Heat approximately 50 cm³ of wine A in a beaker until the volume has been reduced to 1–5 cm³. Cut a strip of chromatography paper approximately 20 cm in length and 1.5–2.0 cm in width. Make a pencil mark approximately 2 cm from one end of the paper. Use the paint brush to transfer the concentrate obtained from wine A to the mark, making several applications on top of each other, drying the paper between each application, and attempting to keep the spot of small size.

Introduce approximately 10 cm³ chromatographic solvent into the Kilner jar. Fix the upper end of the chromatography paper to the side of a 100 cm³ beaker, using plasticine, then invert the beaker over the Kilner jar, as illustrated in Fig. 4. Ensure that one end of the paper dips into the solvent and that the dried solute is above the level of the solvent. Allow the solvent to ascend the paper for 2–4 hours, then remove the paper and dry it. Dip the dried paper into a 100 cm³ measuring cylinder containing 100 cm³ alkaline bromocresol green solution. As a result of this treatment organic acids appear as yellow spots against a blue background. Repeat the procedure with wine B.

Measure the R_f values of succinic, lactic, malic, citric and tartaric acids. Compare these values with those obtained for the two wines, A and B. Record your results and state any conclusions that you are able to draw from this exercise.

Topics to investigate

1 Chemical analyses of beer, lager and cider.
2 Chemical anaylses of spirits.
3 Chemical anaylses of home-brewed wines.
4 Chromatographic separation and identification of organic acids from ripening fruits.

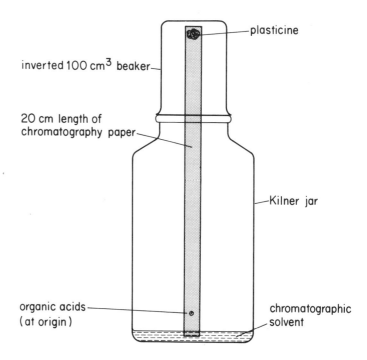

inverted 100 cm³ beaker — plasticine

20 cm length of chromatography paper

Kilner jar

organic acids (at origin)

chromatographic solvent

Fig. 4 *Apparatus for the separation and identification of organic acids in wines*

1.10 Estimation of the chlorophyll and starch content of leaves

A method is described for estimating relative amounts of chlorophyll and starch in individual leaves taken from a potted plant.

Preparation

The iodine solution is prepared by dissolving 2 g iodine crystals and 4 g potassium iodide in 100 cm^3 water.

Materials

- Potted plant of cucumber, bearing 4–8 leaves
- Colorimeter, fitted with a red filter
- 100 cm^3 absolute ethanol
- Two 10 cm^3 plastic syringes
- Iodine solution
- No. 15 cork borer
- Eight boiling tubes, in a rack
- Pestle and mortar
- Water-bath, maintained at 80° C
- Bunsen burner
- Glass marking pen

Method

1 Remove the leaves from the plant, numbering them in order from the shoot apex down the length of the stem to the base.
2 Using the cork borer, cut 15 discs from each leaf, transferring each set of discs to a labelled boiling tube. Add approximately 10 cm^3 water to each tube and boil each set of leaf discs for one minute.
3 Drain the water from each boiling tube and add 10 cm^3 absolute ethanol, marking the position of the meniscus on the outside of the glass.
4 Transfer the boiling tubes, in the rack, to a water-bath maintained at 80° C. Heat the discs until they are white, then remove the tubes from the heat and add ethanol to restore the volume of the solvent to its original level.
5 Transfer some of the solution of chlorophyll from each boiling tube to a colorimeter, fitted with a red filter. Read off the percentage absorbance from the scale. Additionally, measure the absorbance of a blank tube, containing water only. Record your results in the form of a table.
6 Remove the decolourised discs from leaf 1, transfer them to the mortar and grind them to a pulp with 1 cm^3 of the iodine solution. Return the pulp to boiling tube 1, add a further 9 cm^3 iodine solution, mix thoroughly, and transfer some of the supernatant liquid to a colorimeter, fitted with a red filter. Read off the percentage absorbance from the scale. Additionally, measure the absorbance of a blank tube, containing iodine solution only.
7 Repeat procedure 6 with each set of leaf discs. Record your results in the form of a table.
8 Plot all of your results in the form of a histogram, relating the chlorophyll content of each leaf to the starch content.
9 What conclusions do you draw?

Topics to investigate

1 Seasonal changes in the chlorophyll and starch content of leaves.
2 Relationships between light intensity and the chlorophyll and starch content of leaves.
3 The effects of light quality (wavelength) on the development of chlorophyll and starch in leaves of a named plant.
4 Effects of mineral ion deficiencies on the chlorophyll content of leaves.

2 Proteins, Enzymes and Related Compounds

2.1 To determine the rate of denaturation of a protein by heat, ethanol and copper(II) sulphate

TIME 2 h

Egg albumen is denatured by heat and a number of toxins, such as ethanol and copper (II) sulphate. As the egg albumen is denatured, there is an increase in its opacity, which can be measured (as percentage absorbance) in a colorimeter. These measurements can be used as an index of solvent or solute toxicity.

Preparation

Although it is possible to use dried egg albumen in this exercise, the use of albumen from fresh eggs is recommended. The flat-bottom tubes should be of a size that can be accommodated by the colorimeter.

Materials

- Two or three fresh eggs
- 25 cm³ absolute ethanol
- 25 cm³ 0.1 M copper(II) sulphate solution
- 100 cm³ distilled water
- Two 10 cm³ plastic syringes
- Colorimeter
- Thirty flat-bottomed tubes
- Water-baths maintained at 30, 40, 60 and 80°C
- Beaker
- Glass marking pen

Method

1 Crack the egg shells and pour off the albumen into a beaker, retaining the egg yolk in the shell.

2 Introduce approximately 5 cm³ egg albumen into each of sixteen flat-bottomed tubes. Transfer four tubes to the water-bath maintained at 30°C and at intervals of 15 minutes, over a period of one hour, remove one of the tubes and measure the absorbance of the albumen. Similarly, transfer four tubes of albumen to the water baths maintained at 40, 60 and 80°C. Remove a tube of albumen from each water bath every 15 minutes, and measure absorbance. Record your results in the form of a table. Plot your results as a graph and comment briefly on your results.

3 Investigate the effect of ethanol on the denaturation of egg albumen. Introduce 5 cm³ egg albumen and 5 cm³ absolute ethanol (= 50% ethanol by volume in the mixture) into a flat-bottomed tube, gently rock the mixture to ensure that mixing has taken place, and measure absorbance one minute after mixing the two ingredients in the tube. Repeat this procedure with ethanol/water mixtures containing, respectively, 40, 30, 20 and 10% ethanol by volume. Record your results in the form of a table and present your results as a graph. Comment briefly on your results.

4 You are provided with a 0.1 M copper(II) sulphate solution. Devise your own method for determining the lowest concentration of copper(II) sulphate required to bring about complete denaturation of the egg albumen. Write out your method and use your method to obtain a set of results. Present your results in a suitable form and comment briefly on them.

Topics to investigate

1 Effects of pH on the denaturation of egg albumen.
2 Denaturation of egg albumen as an index of (i) solvent and (ii) solute toxicity. (It might be possible to compare the relative toxicity of different alcohols, or a range of solutes such as salts of sodium.)

2.2 Losses of protein and sugars from foods as a result of cooking, freezing and drying

TIME 2 h, April–September

During the storage and preparation of foodstuffs there is often some loss of valuable nutrients, such as proteins and sugars. Semi-quantitative methods are used in this exercise to determine the extent of some of these losses.

Preparation

On seven successive days before the laboratory work is carried out, 1 g samples of fresh grass should be collected and allowed to dry out in a well-ventilated room.

Materials

- Fresh green beans
- Frozen sliced beans
- Fresh carrot
- Frozen carrot
- Fresh grapes (preferably sultana grapes)
- Dried sultanas
- Fresh and dried grass
- Invertase concentrate[1,4]
- Albustix reagent strips[4]
- Dextrostix reagent strips[4]
- Four 250 cm³ beakers
- Pestle and mortar
- Two bunsen burners, gauzes and tripods
- Top-pan balance

Topics to investigate

1 Changes in the protein and sugar content of grass during hay-making and silage-making.

Method

1 Weigh out 5 g fresh beans. Transfer the beans to a mortar, grind them to a paste and measure the protein content of the cell sap, using Albustix reagent strips. Add two drops of invertase concentrate to the slurry, wait for 20 minutes and then measure the glucose content of the material, using Dextrostix reagent strips. (Beans contain sucrose, which must be hydrolysed to a mixture of glucose and fructose before it can be estimated.)

Introduce 20–30 g fresh green beans into a beaker. Add 100 cm³ water and boil the beans over a bunsen burner. At intervals of 15 minutes, over a period of one hour, withdraw a sample of beans, grind them to a pulp and test the slurry for the presence of protein and glucose. At the same time withdraw some of the water in which the beans have been boiled and test this both for protein and for glucose. Record your results in the form of a table.

Measure and record the protein and glucose content of the frozen green beans.

2 Repeat procedure 1 with the fresh carrots. In this case there is no need to add invertase concentrate before testing with Dextrostix. Record your results in the form of a table. Measure and record the protein and glucose content of the frozen carrots.

3 Take three fresh grapes, preferably sultana grapes which are available during August–September, weigh them and transfer them to a mortar. Grind the soft tissues to a pulp and measure the protein and glucose content of the sap. Take three sultanas, weigh them and transfer then to a mortar. Add water to make up the mass to that of the fresh grapes. Measure and record the protein and glucose content of the slurry.

4 Take, in turn, each sample of grass, transfer it to the mortar and grind it to a pulp with 0.5 cm³ water. Measure and record the protein content of each sample. Sucrose is the sugar present in grass so add 1–2 drops of invertase concentrate to each sample, wait for 20 minutes, then measure the glucose content of each sample. Record your results in the form of a table.

5 What conclusions do you draw from all your results? What recommendations, if any, would you make to those concerned with (i) the preparation or preservation of plant material for human consumption, and (ii) the preservation of the grass crop for feeding to cattle?

A method for measuring the density of milk: copper proteinate formation

TIME 1–1½h

A method is described for determining the relative density of protein-containing fluids, such as milk or blood. Measurements are made to determine the rate at which drops of milk, or milk–water mixtures, fall under gravity through a copper(II) sulphate solution. A layer of copper proteinate forms around each drop, preventing dispersal of the milk.

Preparation

Once students have mastered the technique, and wish to investigate the relative density of protein-containing fluids, they may wish to use syringes which deliver drops of uniform size, available from supplier no. 5.

Materials

- 100 cm³ fresh milk
- 100 cm³ distilled water
- 250 cm³ 0.1M copper(II) sulphate solution
- Two 10 cm³ plastic syringes
- 1 cm³ plastic syringe, fitted with a long needle
- Two 100 cm³ measuring cylinders
- Stop-watch

Method

1 Pour the solution of copper(II) sulphate into the two measuring cylinders, filling each one to a depth approximately 5 cm above the 100 cm³ level. Introduce some undiluted milk into the 1 cm³ syringe, fitted with a long needle. Place the tip of the needle just below the surface of the copper(II) sulphate solution. Gently press the plunger of the syringe to release a small drop of milk, and record the time taken for the drop to fall between the 100 cm³ and 10 cm³ marks (see Fig. 5). Repeat this procedure using the second measuring cylinder. Record the time taken for three drops to fall and calculate a mean value.

2 Make dilutions of the milk containing, respectively, 5, 10, 20, 30, 40, 50 and 60% milk by volume. Repeat procedure 1 with each milk–water mixture. Record the time taken for three drops of each mixture to fall between the 100 and 10 cm³ marks, and calculate mean values. Record your results in the form of a table.

3 In what proportion must milk and water be mixed to have a density equivalent to the density of a 0.1M copper(II) sulphate solution? Explain how you arrived at your answer.

4 Plot all your results as a graph.

5 State the chief source of error in this experiment and explain how it could be overcome.

Topics to investigate

1 Variations in the density of different grades of marketed milk.
2 Density differences in milk from different breeds of cattle.
3 A study of condensed milks.
4 Differences in the relative density of blood samples taken from different blood vessels in a rabbit.
5 Variations in the density of human blood in health and disease.
6 Studies of the density of human saliva.

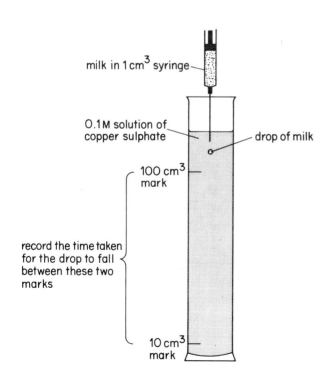

milk in 1 cm³ syringe

0.1M solution of copper sulphate

drop of milk

100 cm³ mark

record the time taken for the drop to fall between these two marks

10 cm³ mark

Fig. 5 *Method for determining the density of milk and milk–water mixtures*

2.4 | Amylase production in germinating grains of barley

During the germination of cereal grains, amylase is synthesised in the aleurone layer of the endosperm under the direct influence of gibberellic acid from the embryo. This exercise demonstrates (i) changes in the level of amylase activity during germination, and (ii) variations in amylase production between the embryo-halves and the endosperm-halves of grains.

Preparation

Seven days before the exercise is presented, set up petri dishes containing 2 g barley sown on moist sawdust, cotton wool or vermiculite. Repeat this procedure on each day preceding the exercise, so that a total of seven dishes, each containing 2 g germinating barley can be presented to each student. Label the dishes A (day one) to G (day seven). While germination is taking place, maintain the dishes at a temperature within the range 20–25 °C.

In part of this exercise separate embryo-halves and endosperm-halves of barley grains are required. Fig. 6 shows the structure of an entire barley grain. Grains should be bisected along the line *a—a* to separate embryo-halves from endosperm-halves. These need to be prepared seven days before the exercise is presented. Place 10 embryo-halves and 10 endosperm-halves in separate stoppered tubes, each batch of half-grains immersed in 2 cm³ water.

The soluble starch solution should be prepared by adding 1 g soluble starch, mixed with a little cold water, to one litre of boiling water. Satisfactory results have been obtained with an iodine solution containing 2 g iodine crystals and 4 g potassium iodide dissolved in 100 cm³ water.

Materials

- Seven batches of germinating barley grains, labelled A to G, sown on successive days
- Ten embryo-halves of barley grains, labelled H
- Ten endosperm-halves of barley grains, labelled I
- Nineteen test-tubes in a rack, or nineteen flat-bottomed tubes
- Nine glass rods
- Two large white tiles
- Pestle and mortar
- 250 cm³ soluble starch solution
- Two 10 cm³ pipettes or 10 cm³ plastic syringes
- Iodine solution
- Fluted filter paper, Whatman no. 1 or 2 grade
- Filter funnel
- Washbottle containing distilled water
- Two 250 cm³ beakers
- Paper towels
- Glass marking pen

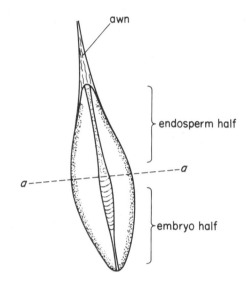

Fig. 6 *An entire barley grain*

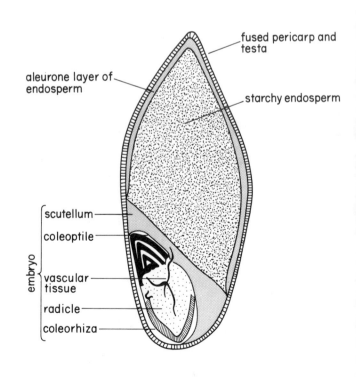

Fig. 7 *A longitudinal section of a barley grain*

Method

1 Grind 40 grains from batch A in a mortar, with $10 \, cm^3$ soluble starch solution. Filter the resulting slurry through fluted filter paper, retaining the filtrate. As filtration may take from 20–30 minutes, you are advised to proceed with other parts of the exercise. The filtrate should contain only soluble starch and amylase, any starch grains from the endosperm being retained by the filter paper. Transfer $1–5 \, cm^3$ of the filtrate to a test-tube or flat-bottomed tube. Immediately withdraw a drop of the mixture from the tube, using a glass rod, and mix it with a drop of iodine solution on a white tile, testing for the presence of starch. Record only if starch is present (+ve) or absent (−ve). Repeat the test at intervals of 8 minutes, over a period of 80 minutes, recording your results in the form of a table.

2 Wash out the mortar and dry it with a paper towel. Repeat the procedures outlined in 1 with grains from dishes B to G. Label each filtrate. Wash and dry the mortar after crushing each batch of grains. Record all your results in the form of a table and present your results as a graph, indicating how the graph was constructed. Comment briefly on your results.

3 Grind up the embryo-halves of grains, labelled H, in their immersion fluid and repeat procedure 1, testing the mixture for the presence of starch. Repeat this procedure with the endosperm-halves of the grains, labelled I. Record your results in the form of a histogram, explaining how the histogram was constructed. How do you account for your results?

Topics to investigate

1 Amylase activity in germinating grains of wheat, maize and oat.
2 Amylase activity in germinating legumes, with particular reference to amylase production by the testa.
3 Amylase activity in white and brown flour.
4 The effect of gibberellin on amylase production in germinating barley grains.

Variations in the rate of salivary amylase activity

Starch-impregnated paper discs are used in this investigation of salivary amylase activity. After investigating amylase activity in your own saliva, you are asked to compare this with salivary amylase activity in other members of your class or set.

Preparation

Paper discs should be cut from starched paper using a paper punch. Satisfactory results have been obtained with Roneo duplicating paper, certain types of newspaper, bank paper, typing paper and carbon-copying paper. Even so, it is advisable to conduct trials before a paper is used in the experiment. Use the most dilute iodine solution that gives a blue colour when applied to the starched paper.

Materials

- Twenty to thirty paper discs
- 9 cm petri dish
- Flat-bottomed tube, containing iodine solution
- 10 cm³ plastic syringe
- White tile
- Forceps
- Paper cup, containing a pinch of sodium chloride dissolved in 10 cm³ water
- Oven, maintained at 35 °C

Method

1 Obtain a sample of saliva by taking the 10 cm³ saline solution into the buccal cavity and chewing it over exactly 50 times before expelling it into the cup.
2 Using the 10 cm³ syringe, introduce 10 cm³ saliva into the petri dish. Add 20–30 paper discs to the saliva, separating the discs from one another. Replace the lid of the dish and transfer it to the oven.
3 At intervals of 5 minutes, over a period of up to 60 minutes, use the forceps to remove a paper disc from the saliva. Dip the paper disc into the iodine solution for 10 seconds, then transfer the disc to the white tile. If necessary, repeat the test with a second disc.
4 Observe any colour changes in the paper discs that occur during the experiment. Record your results either as a table or as a drawing.
5 Present your results, and those from other members of your class, in the form of a table. Construct a histogram from the table, and write a brief discussion of variations in amylase activity.
6 Describe the basis of the experiment you have performed. Evaluate the accuracy of the method compared with the more usual technique of mixing together drops of iodine solution and a suspension of starch grains on a white tile.

Topics to investigate

1 A comparative study of methods used to determine salivary amylase activity.
2 Effects of salivary amylase on the rate of hydrolysis of starches obtained from (i) potato, (ii) wheat and (iii) rice.
3 Variations in salivary amylase activity with age, sex and diet.
4 Does smoking affect salivary amylase activity?
5 Amylase activity in potato cell sap.

2.6 The effects of α-amylase and β-amylase on the digestion of starch

TIME $3\frac{1}{2}$ h

Amylase powders, purchased from suppliers, generally consist of two components, α-amylase and β-amylase. The purpose of this investigation is to compare the rates at which starch is hydrolysed by α-amylase and β-amylase, used separately and in combination.

Preparation

α-amylase and β-amylase are available from supplier no. 1. Make up solutions of each enzyme by dissolving 0.5 g in 100 cm³ water.

Materials

- Three 40 cm lengths of 1.5 cm diameter Visking tubing
- 50 cm³ α-amylase solution
- 50 cm³ β-amylase solution
- 50 cm³ mixed α-amylase and β-amylase (25 cm³ of each)
- Three 10 g samples of starch
- Water-bath, maintained at 35°C
- Top-pan balance
- String
- Glass marking pen

Method

1 Take three lengths of Visking tubing, number them and tie off one end of each tube. Introduce 10 g starch into each tube. Make further additions as follows:

   Tube 1 50 cm³ α-amylase solution;
   Tube 2 50 cm³ β-amylase solution;
   Tube 3 50 cm³ mixed α-amylase and β-amylase solution.

   Seal the open end of each piece of tubing with string, leaving sufficient head room for the mixture inside the tube to increase in volume.

2 Transfer the three sealed tubes to a water-bath maintained at 35°C. After immersion immediately remove the tubes, dry them on the outside and record the mass of each tube. Re-weigh them at intervals of 30 minutes over a period of 3 hours. Record your results in the form of a table and present your results as a graph.

3 Explain (i) the principle on which the experiment is based, and (ii) the effects of α-amylase and β-amylase on the hydrolysis of starch.

Topics to investigate

1 Chemical inhibition of α-amylase and β-amylase. In view of the high cost of amylase, an alternative form of this experiment can be carried out using mixed α-amylase and β-amylase. The action of α-amylase can be inhibited by the use of sodium hexametaphosphate and β-amylase by the use of copper (II) sulphate.

2 Do saliva and pancreatic juice contain α-amylase and β-amylase?

2.7 The hydrolysis of sucrose by invertase

TIME 2–2½ h

The enzyme invertase catalyses the reaction

$$Sucrose \longrightarrow Fructose + Glucose$$

Sucrose	Fructose	Glucose
(*Non-reducing sugar*)	(*Reducing sugar*)	(*Reducing sugar*)

As the products of the reaction are both reducing sugars, the procedure described in exercise 1.4 can be used to monitor the rate of the reaction.

Preparation

The sucrose solution contains 10 g sucrose dissolved in 100 cm³ water.

Materials

- Invertase concentrate[1]
- 100 cm³ sucrose solution
- 20 cm³ 0.01M potassium manganate (VII) (potassium permanganate)
- 50 cm³ 1M sulphuric acid
- two 10 cm³ plastic syringes
- nine flat-bottomed tubes, each approximately 2 × 8 cm

Method

1 Follow the procedure outlined in exercise 1.4 to produce a standard calibration curve for concentrations of glucose in solution ranging from 1–10 g per 100 cm³.
2 Add 2 cm³ invertase concentrate to 100 cm³ of the sucrose solution, and stir the mixture. One minute after making this addition, and again after 5, 10, 15, 20, 25, 30, 60 and 90 minutes, measure the amount of reducing sugar present in the sucrose–invertase mixture. Use the same volume of each component as in exercise 1.4, namely: 10 cm³ sugar solution, 5 cm³ sulphuric acid and 2 cm³ potassium manganate (VII) solution. Record the time taken for the permanganate to become decolourised in each test. Present your results in the form of a table, estimate amounts of reducing sugars present, using the standard curve, and plot your results as a graph.
3 Comment briefly on your results.

Topics to investigate

1 Effects of temperature on the hydrolysis of sucrose by invertase.
2 Effects of acids on the hydrolysis of sucrose.
3 Hydrolysis of maltose by maltase (β-galactosidase)
4 Invertase activity of soil samples.
5 Invertase activity of yeasts.

2.8 To determine the rate at which milk is coagulated by rennin at room temperature

TIME 2 h

The enzyme rennin is found in the stomach of young mammals, where it solidifies milk by converting caseinogen into casein, which is precipitated as calcium caseinate.

$$\text{Caseinogen} + Ca^{2+} \longrightarrow \text{Calcium caseinate}$$

When rennin is added to fresh milk the viscosity of the mixture increases, and changes in the flow rate can be used to monitor the course of the reaction.

Preparation

Junket tablets are generally available from food and health stores.

The most suitable temperature for this experiment is 15–18°C. If the experiment is carried out at higher temperatures, the milk gels too rapidly, making it difficult to obtain meaningful results.

Materials

- 150 cm^3 fresh milk in a beaker
- 250 cm^3 beaker
- Junket tablet
- Pestle and mortar
- Ten 10 cm^3 plastic syringes
- Boss and clamp
- Test-tube rack (to hold syringes)
- Glass stirring rod
- Stop-watch
- Ruler

Method

1 Crush the junket tablet in the mortar and add the powder to the milk, stirring the mixture to ensure that the enzyme is evenly distributed.
2 Fill each of the ten 10 cm^3 plastic syringes with the mixture to the 10 cm^3 mark, then support the filled syringes in the test-tube rack, with the barrel of each syringe inserted through one of the holes in the rack. Projections at the top of each syringe barrel should hold the syringes in position.
3 Immediately, and at intervals of 10 minutes, over a period of 90 minutes, gently and carefully lift one of the syringes from the rack. Support it, held vertically in the clamp, with the nozzle positioned approximately 10 cm above the bench surface. Place the empty beaker beneath the nozzle. Gently lift out the plunger from the barrel, recording the time taken for the milk to drain, under gravity, through the nozzle of the syringe. Repeat this procedure with each of the milk–rennin mixtures. Record your results in the form of a table and plot your results as a graph.
4 Comment briefly on the procedures employed and the results obtained.
5 What other enzyme-catalysed reactions could be monitored by measuring changes in the viscosity of the end product? Name specific examples involving (i) increased viscosity of the product as the reaction proceeds, and (ii) reduced viscosity of the product as the reaction proceeds.

Topics to investigate

1 Effects of temperature on the coagulation of milk by rennin.
2 Coagulation of milk by pepsin.
3 Factors affecting the rate at which blood coagulates.
4 Experiments involving anti-coagulants.

2.9 The effect of temperature on the rate of an enzyme-catalysed reaction

TIME 1–2 h

Powdered milks contain a white protein called casein. A white, opaque suspension of Marvel milk in water loses its opaqueness and becomes translucent following hydrolysis by proteolytic enzymes. Providing the surrounding medium is kept alkaline, and an enzyme such as trypsin is used, Marvel milk is a convenient substrate for demonstrating the effect of temperature on an enzyme-catalysed reaction. (It should be noted, however, that in acid solutions Marvel milk undergoes rapid hydrolysis and dissolves, even without the addition of an enzyme.)

Preparation

The enzyme substrate is prepared by adding 4 g Marvel milk to 100 cm³ water. The enzyme solution contains 0.5 g trypsin[1] dissolved in 100 cm³ water.

Materials

balance
Biolog.
washing
powder

- 50 cm³ Marvel milk suspension
- 40 cm³ trypsin solution
- 10 cm³ 0.1 M hydrochloric acid
- 10 cm³ distilled water
- Thirteen flat-bottomed tubes, each approximately 2 × 8 cm
- Two 10 cm³ pipettes, or plastic syringes
- Water-baths maintained at
  (i) room temperature + 10 °C;
  (ii) room temperature + 20 °C;
  (iii) room temperature + 30 °C;
  (iv) room temperature + 40 °C;
  (v) room temperature + 50 °C.

Method

1 Introduce 5 cm³ Marvel milk suspension and 5 cm³ distilled water into a flat-bottomed tube. This is a colour standard, which may be used to indicate the absence of any enzyme activity.

2 Introduce 5 cm³ Marvel milk suspension and 5 cm³ hydrochloric acid into a flat-bottomed tube. What changes do you observe and how do you account for these changes?

This is a second colour standard, which may be used to determine the point at which digestion of the enzyme substrate is complete.

3 Introduce 5 cm³ Marvel milk suspension and 5 cm³ trypsin solution into a flat-bottomed tube. Leave the mixture, at room temperature, on the bench surface and record the time taken for complete digestion of the Marvel milk.

4 Introduce 5 cm³ Marvel milk suspension into each of five flat-bottomed tubes. Introduce 5 cm³ trypsin solution into each of five other flat-bottomed tubes. Transfer one tube containing Marvel and one containing trypsin solution to a water-bath at 10 °C above room temperature. Allow approximately five minutes for the contents of each tube to reach the temperature of the water-bath, then mix the contents of the two tubes, maintaining the mixture at the same temperature. Observe the mixture for hydrolysis of the Marvel milk and record the time taken for the mixture to clear.

Similarly, transfer one tube containing Marvel milk and one containing trypsin solution to each of the other water baths. Again, record the time taken for complete hydrolysis of the enzyme substrate. Record all your results in the form of a table.

5 Present your results as a graph. Comment briefly on your results.

6 Propose modifications in technique or procedure which would improve the accuracy or validity of the results obtained.

Topics to investigate

1 Optimal temperatures for the activity of different enzymes, including those from 'biological' washing powders.

2.10 To determine the effect of temperature on the inactivation of trypsin

TIME 1–2 h

The same enzyme and enzyme substrate are used in this exercise as in exercise 2.9. In this case, the object is to demonstrate the effect of temperature on the denaturation of the enzyme.

Preparation

The enzyme substrate is prepared by adding 1 g Marvel milk to 100 cm³ water. The enzyme is prepared by dissolving 0.5 g trypsin[1] in 100 cm³ water.

Materials

- 100 cm³ Marvel milk suspension
- 100 cm³ trypsin solution
- Thirteen flat-bottomed tubes, approximately 2 × 8 cm
- Two 250 cm³ beakers
- Two 100 cm³ beakers
- Two 10 cm³ pipettes, or plastic syringes
- Water-bath maintained at 60° C
- Water-bath maintained at 80° C
- Thermometer

Method

1 Introduce 5 cm³ Marvel milk suspension and 5 cm³ trypsin solution into a flat-bottomed tube. Stand the tube on the bench surface at room temperature and record the time taken for completion of the reaction, which is indicated by the mixture becoming translucent and colourless (see exercise 2.9).

2 Introduce 50 cm³ trypsin solution into a 250 cm³ beaker and transfer it to the water-bath maintained at 60° C. Similarly, transfer a second beaker containing 50 cm³ trypsin solution to the water-bath maintained at 80° C. At intervals of five minutes over a period of 30 minutes pour some of the trypsin maintained at 60° C into a 100 cm³ beaker. Using a pipette, or plastic syringe, transfer 5 cm³ of the trypsin solution to a flat-bottomed tube. Return any surplus of trypsin to the beaker at 60° C. Allow the trypsin in the tube to cool to room temperature.

   Similarly, at intervals of five minutes over a period of 30 minutes pipette off 5 cm³ samples of the trypsin solution maintained at 80° C. Allow these samples to cool to room temperature.

3 When all the samples of trypsin have cooled to room temperature, add 5 cm³ Marvel milk suspension to each tube. Record the time taken for completion of the reaction in each tube, presenting your results in the form of a table.

4 Plot your results as a graph to show the relationship between percentage activity of the enzyme (the unheated enzyme shows 100% activity) and duration of the heat treatment, both at 60° C and 80° C.

5 From your results suggest any modifications of the experiment, or further experiments you would wish to carry out, in order to examine the effect of heat treatment on the activity of trypsin.

Topics to investigate

1 Effects of (i) low temperatures and (ii) high temperatures on the inactivation of enzymes.
2 Deterioration of enzymes as a result of storage.

2.11 To determine the optimal pH for the activity of two enzymes

TIME 1½ (urease), 4 h (zymase)

You are asked to investigate the effects of pH on the activity of two enzyme-catalysed reactions.

Urease catalyses the reaction:

$$Urea \rightarrow NH_3 + CO_2$$

Zymase is the name given to a complex of enzymes in yeast responsible for the fermentation of glucose:

$$C_6H_{12}O_6 \rightarrow 2C_2H_5OH + 2CO_2$$

Preparation

Buffer solutions are most conveniently prepared from tablets. Dissolve one tablet, intended to make 100 cm³ buffer, in 50 cm³ water. Each student will require 100 cm³ of each buffer. The suspension of yeast contains 10 g dried yeast added to 100 cm³ water. Solutions of urea and glucose each contain 10 g solute dissolved in 100 cm³ water.

Materials

- Buffer solutions (pH 4.0, 6.4, 6.8, 7.4, 8.0 and 9.0)
- 100 cm³ urea solution
- Urease tablets[1]
- Seven small beakers
- Three 10 cm³ pipettes, or plastic syringes
- Merckoquant ammonia-sensitive reagent sticks[1]
- 50 cm³ yeast suspension
- 50 cm³ glucose solution
- Twelve 10 cm³ plastic syringes
- Six bosses and clamps
- Rubber tubing (to fit over nozzle of syringes)
- Glass marking pen

Method

1 Dissolve 5 urease tablets in the urea solution. Pipette 10 cm³ of the urea–urease mixture into each of six numbered beakers. Add 10 cm³ pH 4.0 buffer to the first beaker and add the same volume of the other buffers to beakers 2–6. Allow the mixtures to stand on the bench surface at room temperature.

2 After 15 minutes, and again after 60 minutes, dip a Merckoquant ammonia-sensitive reagent stick into each of the buffered mixtures, following the manufacturer's instructions. Read off the concentrations of ammonia, and record your results in the form of a table. Plot your results as a graph and comment briefly on your results.

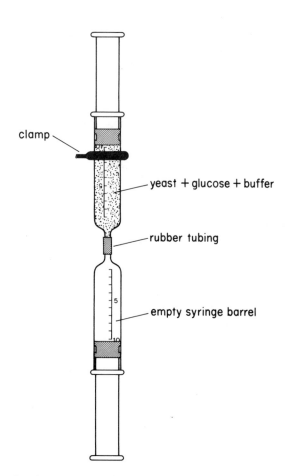

Fig. 8 *Arrangement of the syringes*

clamp

yeast + glucose + buffer

rubber tubing

empty syringe barrel

3 Mix the suspension of yeasts with the glucose solution. Draw 5 cm³ of the mixture into each of six numbered 10 cm³ syringes. Fill each syringe to the 10 cm³ mark by introducing 5 cm³ of the appropriate buffer, pH 4.0–9.0. Gently rock the contents of each syringe to ensure that mixing has taken place.

4 Take an empty 10 cm³ syringe, pull out the plunger to the 10 cm³ mark, and use rubber tubing to attach the nozzle of the syringe to a syringe containing a mixture of yeast, glucose and buffer (see Fig. 8). Support the two joined syringes in a clamp. Do this for each of the six different buffer solutions.

5 After 3–4 hours measure either (i) the volume of the mixture displaced from the upper syringe, or (ii) the volume of the mixture collected in the lower syringe. Record your results in the form of a table. Plot your results as a graph, and comment briefly on your results.

Topics to investigate

1 Optimal pH for (i) amylases and (ii) lipases from different sources.

The effect of pH on the activity of two proteolytic enzymes   TIME 2–2½ h

You are asked to investigate the effect of pH on the activity of two proteolytic enzymes, pepsin and trypsin. The substrate for the enzymes is a block of table jelly, which consists principally of the protein gelatine.

Preparation

Prepare the buffers as in exercise 2.11. Each student will require 100 cm³ of each buffer. Solutions of both pepsin and trypsin, which need to be fairly concentrated if results are to be obtained in 2–3 hours, may be prepared by dissolving 1 g enzyme powder[1] in 100 cm³ water. It is advisable to use a table jelly which is highly coloured, such as the blackcurrant or raspberry flavours.

Materials

- Block of table jelly
- Ten 9 cm petri dishes
- Buffer solutions (pH 4.0, 6.4, 7.4, 8.0 and 9.0)
- 50 cm³ pepsin solution, labelled A
- 50 cm³ trypsin solution, labelled B
- Two 10 cm³ pipettes, or plastic syringes
- Sharp knife or scalpel
- Glass marking pen
- Graph paper, graduated in millimetre and centimetre squares.

Method

1 Using the sharp knife, cut strips of jelly from the block, each of approximately 1–2 mm in thickness. Transfer one of these strips to each petri dish. Place a piece of graph paper under each dish and trim each strip of jelly to a rectangle of 1 × 2 cm; discard any surplus jelly.
2 Introduce 10 cm³ of each buffer into two of the petri dishes, using the pipette or plastic syringe. Label each dish. Divide the dishes into two sets.
3 Add to one set of dishes 10 cm³ of enzyme A and to the other set 10 cm³ of enzyme B. Label the dishes.
4 Allow the two sets of dishes to stand on the bench surface at room temperature. At intervals of 15 minutes, over a period of 90 minutes, measure the area of the gelatine strip in each dish. Record your results in the form of a table. Plot your results as a graph. What do you deduce about the activity of the two enzymes?
5 Offer five or more criticisms of the experimental procedures used in this exercise.

Topics to investigate

1 Proteolytic enzyme activity and pH in the alimentary canal of (i) a locust and (ii) an earthworm.
2 Comparisons between the activity of pepsin and trypsin.

The effect of pH on phosphatase activity in the testis and epididymis of a male rat

TIME 2–2½ h

The reaction

ADP	+	P		ATP
Adenosine diphosphate		*Inorganic phosphate*		*Adenosine triphosphate*

is catalysed by enzymes called phosphatases. Active tissues may contain several different phosphatases, each acting at a different pH.

If the artificial enzyme substrate phenolphthalein diphosphate is acted upon by phosphatases, the following reaction occurs:

Phenolphthalein diphosphate $\longrightarrow$ Phenolphthalein + Phosphate
(*No reaction with alkali*) (*Reacts with alkali to form a pink–purple chromogen*)

Addition of an alkaline solution, such as sodium carbonate, to the products of the reaction results in the development of a pink colouration, the intensity of the colour being directly related to the amount of enzyme present. The object of this exercise is to investigate phosphatase activity in the testis and epididymis of a rat.

Preparation

Again, as in the previous exercises, the buffers are most conveniently prepared from tablets by dissolving one tablet in 50 cm³ water. Each student will require 50 cm³ of each buffer. Prepare the enzyme substrate by dissolving 0.5 g of the sodium or calcium salt of phenolphthalein diphosphate[1] in 100 cm³ water. A slight cloudiness of the solution will not affect the results. Owing to the cost of rats, a single large male could be used to provide material for from 5–10 students.

Materials

- Freshly-killed or deep-frozen mature male rat
- Dissecting board, dissecting instruments, awls, pins
- Buffer solutions (pH 4.0, 6.4, 7.4, 8.4 and 9.4)
- 60 cm³ phenolphthalein diphosphate solution
- 20 cm³ 0.5 M sodium carbonate solution
- 5 cm³ 0.05 M sodium carbonate solution
- Phenolphthalein solution (indicator)
- 5 cm³ plastic syringe
- 1 cm³ plastic syringe
- Pestle and mortar
- Twenty test-tubes in a rack, or twenty flat-bottomed tubes
- Small beaker, containing clean, dry sand

Method

1 Dissect out a single testis and epididymis from the rat provided and transfer the dissected material to a mortar. Gently macerate the tissue in the mortar, grinding it, if necessary, with a little clean, dry sand. Add 50 cm³ phenolphthalein diphosphate solution and continue the grinding process for 1–2 minutes.

2 Using the 5 cm³ syringe as a pipette, transfer 5 cm³ of tissue extract to each of ten labelled tubes. Buffer the tubes, in duplicate, by adding 5 cm³ of a buffered solution. Label the tubes. Allow the mixtures to stand on the bench surface for at least one hour.

3 Prepare a set of colour standards by introducing 1 cm³ phenolphthalein indicator into each of ten test-tubes or flat-bottomed tubes. Number the tubes from 1 to 10. Add a single drop of 0.05 M sodium carbonate solution to the first tube. Increase the number of drops added to each of the remaining tubes, so that each numbered tube contains the same number of drops of sodium carbonate solution. This should produce a range of colours from almost colourless (1 drop), through pale pink (2–6 drops), to deep pink–purple (10 drops).

4 Return to the tissue extract–phenolphthalein diphosphate mixtures. After they have stood for at least one hour, use the 1 cm³ syringe to add a single drop of 0.5 M sodium carbonate solution to each tube. Observe any colour change and the intensity of the colour developed in each tube. If no colour develops, or if the colour is faint, add a second drop of sodium carbonate solution to each tube. Repeat these additions until you can observe an obvious difference in the intensity of the pigment in each tube, matching, as far as possible, the intensity of your colour standards.

5 Estimate the colour intensities in each tube on a scale from 1 (colourless) to 10 (deep pink–purple). Record your results in the form of a table and plot them as a graph. How do you account for your results?

6 Comment on and criticise the experimental procedures employed in this exercise and, wherever possible, propose improvements.

Topics to investigate

1 Phosphatase activity in the organ systems of a mammal.
2 Phosphatase activity in germinating seeds.
3 Phosphatase activity in soil samples.
4 Phosphatase activity in milk.
5 Artificial enzyme substrates.

2.14 The effects of pH and lysozyme on the rate of sedimentation of an aqueous suspension of yeast

TIME $1\frac{1}{2}$–2 h

The enzyme lysozyme, present in tears, can cause the precipitation of certain microorganisms, including yeast and bacteria. The object of this exercise is to demonstrate that rates of precipitation are influenced by pH, and to determine rates of precipitation in different buffered solutions.

Preparation

Buffers are most conveniently prepared from tablets. Dissolve one tablet, intended for the preparation of $100 \, cm^3$ solution, in $50 \, cm^3$ water. Each student will require $50 \, cm^3$ of each buffer. Dissolve 0.5 g lysozyme[1] in $100 \, cm^3$ water. Prepare the suspension of yeast by adding 10 g dried yeast to $100 \, cm^3$ water.

Materials

- Five $100 \, cm^3$ measuring cylinders
- $250 \, cm^3$ yeast suspension in a beaker
- Buffer solutions (pH 4.0, 6.4, 7.0, 8.0 and 9.0)
- $25 \, cm^3$ lysozyme solution
- Ruler, graduated in millimetres
- Stirring rod.

Method

1 Pour $50 \, cm^3$ of each buffer into each of five measuring cylinders. After thoroughly stirring the suspension of yeast with the stirring rod, pour $50 \, cm^3$ of the suspension into each measuring cylinder. Gently rock the contents of each vessel to ensure mixing, taking care to avoid any spillages.

2 Allow the mixtures to stand on the bench surface for 20 minutes, then measure (i) the depth of sediment at the bottom of each cylinder, and (ii) the depth of any yeast-free, clear liquid at the top of the suspension. Record your results. What, if any, is the effect of pH on the rate at which yeast is sedimented from an aqueous suspension?

3 Add to each measuring cylinder $5 \, cm^3$ of the lysozyme solution. Gently rock each vessel to ensure that mixing has taken place, again taking care to avoid any spillages. Allow the mixtures to stand on the bench surface for 20–30 minutes, then measure, as before: (i) the depth of sediment at the bottom of each cylinder, and (ii) the depth of any yeast-free, clear liquid at the top of the suspension. Record your results. What effect, if any, does lysozyme have on the rate at which yeast is sedimented from an aqueous suspension?

4 Plot all your results as a graph.

5 Outline any further investigations you would undertake in order to obtain a better understanding of the effects of pH and lysozyme on the rate of sedimentation of yeast from an aqueous suspension.

Topics to investigate

1 Effects of lysozyme on rates of sedimentation of a named bacterium.

2 Effects of (i) enzymes and (ii) electrolytes on rates of sedimentation of blood corpuscles.

3 Effects of electrolytes on rates of sedimentation of clay particles.

4 Antibody–antigen reactions.

The effect of ascorbic acid (vitamin C) on levels of diphenol oxidase activity

Apples and potatoes contain an oxidase, diphenol oxidase, which oxidises certain colourless phenols to coloured oxidation products:

Phenol + O_2 $\longrightarrow$ Oxidised phenol
(*Colourless*)   (*Brown/black*)

Additionally, apples and potatoes contain ascorbic acid (vitamin C), a powerful reducing agent, which has an inhibitory effect on the oxidase. Using the reagents provided, investigate some of the properties of diphenol oxidase.

Preparation

Prepare the 2,6-dichlorophenol indophenol solution from tablets[1] by dissolving 10 tablets in 100 cm³ water. The catechol solution contains 1 g/100 cm³ water, and the ascorbic acid solution 2 g/100 cm³ water.

Materials

- Apple
- Potato
- 10 cm³ ascorbic acid solution
- 10 cm³ 5-volume hydrogen peroxide solution
- 30 cm³ catechol solution
- 50 cm³ standardised solution of 2,6-dichlorophenol indophenol (DCPIP)
- 50 cm³ distilled water
- Burette or 10 cm³ plastic syringe
- Pestle and mortar
- Cork borer (no. 12–16)
- Knife
- Ruler
- Eight flat-bottomed tubes, each approximately 2 × 8 cm
- Two small beakers.

Topics to investigate

1 Effects of diphenol oxidase on different phenolic substrates. (Diphenol oxidase is an enzyme capable of acting on a number of different substrates.)
2 Effects of (i) pH and (ii) temperature on the activity of diphenol oxidase.
3 Diphenol oxidase activity in fruits and vegetables.
4 Inhibition of diphenol oxidase by (i) uric acid, (ii) cysteine and (iii) chloride ions. (Each of the solutes listed has an inhibitory effect on the activity of diphenol oxidase.)

Method

1 Cut small segments from the apple and potato and place these on the bench surface, with the cut surfaces exposed to air. Leave the segments for approximately one hour, then describe any changes that have occurred, and explain the economic effect of these changes.
2 Estimate the vitamin C content of the apple by using the standardised solution of dichlorophenol indophenol which, on reduction by vitamin C, undergoes a colour change from blue to colourless.

    Blue DCPIP $\xrightarrow{\text{Reduction by vitamin C}}$ Colourless DCPIP

The DCPIP solution has been prepared by dissolving tablets in water. Each tablet, which has a titration equivalent of 1 mg ascorbic acid, has been dissolved in 10 cm³ water.

Using the cork borer provided, cut a solid rod of tissue, 2–3 cm in length, from the apple. Weigh the rod, transfer it to the mortar, add approximately 5 cm³ distilled water and grind the tissue to a slurry. Transfer the slurry to a beaker and titrate the slurry against DCPIP solution, contained either in a burette or plastic syringe. The end point is reached when the blue colour persists. Record your results.
3 Repeat procedure 2 with the potato. Record your results.
4 Investigate phenol oxidase activity in the apple and potato, using the solution of catechol as an artificial substrate for the enzyme. Use the cork borer provided to cut three rods of tissue, each 2–3 cm in length, from the apple. Transfer the rods to the mortar, add 5–10 cm³ distilled water and grind the tissue to a slurry. Subdivide the slurry between four flat-bottomed tubes, and make the following additions to each tube:

Tube 1   10 cm³ distilled water
Tube 2   5 cm³ catechol solution + 5 cm³ distilled water
Tube 3   5 cm³ catechol solution + 5 cm³ ascorbic acid solution
Tube 4   5 cm³ catechol solution + 5 cm³ hydrogen peroxide solution

Allow the tubes to stand on the bench surface for 30 minutes. Record your observations and attempt an explanation. Apply the same procedure to rods of potato tissue. Record your observations and attempt an explanation.
5 What general conclusions do you draw from this investigation?

The effect of concentration of (i) enzyme and (ii) enzyme substrate on the rate of an enzyme-catalysed reaction

TIME $1\frac{1}{2}$–2 h

The enzyme catalase is present in dried yeast. The enzyme substrate is hydrogen peroxide, supplied as a 5-volume solution.

The reaction may be summarised by the equation:

$$2H_2O_2 \longrightarrow 2H_2O + O_2$$

The apparatus used in this exercise can be used in other investigations in which one or more of the products of the reaction is a gas.

Preparation

The apparatus, illustrated in Fig. 8, is assembled from a boiling tube, rubber bung, 1 cm³ or 10 cm³ plastic syringe and needle, a 30 cm length of glass tubing, the barrel of a 20 cm³ syringe, rubber tubing and a screw clip. Two bosses and clamps are required to support various parts of the apparatus in their correct positions.

Prepare the suspension of yeast by adding 10 g dried yeast to 100 cm³ water, at least one hour before the suspension is required for use.

Materials

- Apparatus, as illustrated in Fig. 9
- 200 cm³ yeast suspension
- 50 cm³ 5-volume hydrogen peroxide solution
- Three 10 cm³ plastic syringes
- Glass stirring rod
- Access to water taps and a sink.

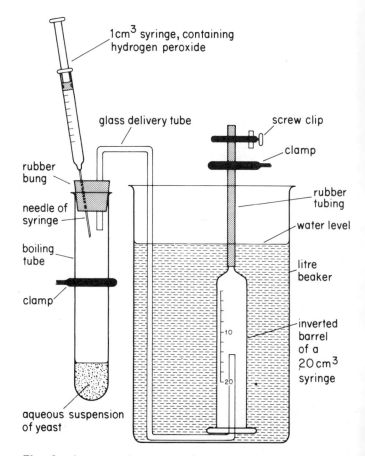

Fig. 9 *Apparatus for exercise 2.16*

Labels: 1cm³ syringe, containing hydrogen peroxide; glass delivery tube; screw clip; clamp; rubber bung; rubber tubing; needle of syringe; water level; boiling tube; litre beaker; clamp; inverted barrel of a 20 cm³ syringe; aqueous suspension of yeast

Method

1 Set up the apparatus as illustrated in Fig. 9, using clamps to support the boiling tube and the rubber tubing attached to the barrel of the 20 cm³ syringe.
2 Remove the 1 cm³ plastic syringe, leaving the needle in position, then remove the bung from the boiling tube. After stirring the yeast suspension, use a plastic syringe to introduce 5 cm³ of the suspension into the boiling tube. Fill the 1 cm³ syringe with the solution of hydrogen peroxide, and place it into position, as illustrated.

 WARNING: Hydrogen peroxide may cause burns to skin or clothing. Wash off any spillages, using plenty of water.

Open the screw clip to draw water into the barrel of the 20 cm³ syringe; close the clip when the barrel is full.
3 Depress the plunger of the 1 cm³ syringe to inject the hydrogen peroxide solution into the boiling tube. Measure and record the volume of oxygen collected in the barrel of the 20 cm³ syringe over a period of five minutes.

Repeat this procedure with a fresh sample of yeast and hydrogen peroxide. Record, at intervals of one minute, the volume of gas collected. Record your results as a histogram. Comment briefly on your results.
4 Using 10, 15, 20, 25 and 30 cm³ yeast suspension in the boiling tube, repeat procedures outlined in 2 and 3, recording your results in the form of a table.
5 Plot your results as a graph. Comment on the shape of the graph.
6 Devise your own method for determining the effect of concentration of enzyme substrate (hydrogen peroxide) on the rate of the reaction. It is recommended that you use one of the 10 cm³ syringes in place of the 1 cm³ syringe as a container for hydrogen peroxide. Describe your method, use it to obtain a set of results, plot your results as a graph, and comment on the shape of the graph.

Topics to investigate

1 Effects of carbonic anhydrase on release of CO_2 from an acidified solution of sodium carbonate.
2 Effects of urease on release of gaseous products from a solution of urea.
3 Optimal substrate concentrations for different enzymes.

Nitrate reductase and nitrite reductase are enzymes that play a role in nitrogen assimilation in plants.

Nitrate reductase catalyses the reduction of nitrates to nitrites:

$$2XNO_3 \xrightarrow{\text{Nitrate reductase}} 2XNO_2 + O_2$$

Nitrite reductase reduces nitrite to ammonia, using ferredoxin as a source of electrons:

$$XNO_2 \xrightarrow{\text{Nitrite reductase}} XNH_4$$

Using the reagents provided, investigate the distribution and activity of these enzymes in the leaves, stems and roots of the bean seedlings.

Preparation

Bean seedlings should be germinated in a nitrogen-rich soil for 5–8 weeks. Each seedling should be washed before it is used in the experiment. Solutions of sodium nitrite and sodium nitrate each contain 1 g solute dissolved in 100 cm³ water.

Materials

- Six seedlings of dwarf French bean (*Phaseolus vulgaris*)
- 30 cm³ sodium nitrite solution
- 30 cm³ sodium nitrate solution
- Six flat-bottomed tubes, each approximately 2 × 8 cm
- Pestle and mortar
- Two 10 cm³ plastic syringes
- Merckoquant nitrite-sensitive reagent sticks[1]
- Merckoquant nitrate-sensitive reagent sticks[1]
- Bench lamp, fitted with a 60 W bulb
- Top-pan balance

Method

1 Sub-divide three bean seedlings into leaves, stems and roots. Weigh 2 g of leaf material, transfer it to the mortar, add 10 cm³ sodium nitrate solution and grind to a slurry. Transfer the slurry to a flat-bottomed tube. Repeat this procedure with the stems and roots of the seedlings.
2 At intervals of 15 minutes, over a period of 1½ hours, dip a Merckoquant nitrite-sensitive reagent stick into each tube. Record concentrations of nitrite in the form of a table. Plot your results as a graph. Comment briefly on your results.
3 Repeat procedure 1 using 10 cm³ sodium nitrite solution and the three remaining bean seedlings. Transfer each slurry to a flat-bottomed tube and stand the tubes at a distance of 5–10 cm from the illuminated bench lamp.
4 At intervals of 15 minutes, over a period of 1½ hours, dip Merckoquant ammonia-sensitive reagent sticks into each tube. Record concentrations of ammonia in the form of a table, and plot your results as a graph. Comment briefly on your results.
5 Outline briefly the role of nitrate reductase and nitrite reductase in the assimilation of nitrogen. Do you consider that this exercise could be put to practical use?

Topics to investigate

1 Location of (i) nitrate reductase and (ii) nitrite reductase activity in plants of different species.
2 Nitrate reductase activity in plants as an index of the nitrogen status of soils.
3 Nitrite reductase activity in plants as an index of the nitrogen status of soils.

2.18 Enzyme induction in seeds of mung bean

TIME 6–24 h

Certain enzymes, called inducible enzymes, are not produced continuously throughout the life of an organism, but only when the chemical inducer of the enzyme becomes available. Germinating seeds of mung bean, if supplied with nitrate and sucrose in solution, are capable of synthesising enzymes, nitrate reductase and invertase respectively. These enzymes can break down the substrates into simpler compounds that the germinating seeds can utilise.

This exercise involves an investigation of the rate of enzyme induction in mung beans.

Preparation

Viable mung beans are generally available from food stores. The sodium nitrate solution, for the induction of nitrate reductase, contains $1 g/100 cm^3$ water, and the sucrose solution, for the induction of invertase, contains $5 g/100 cm^3$ water.

Materials

- Soaked mung beans
- $10 cm^3$ sodium nitrate solution
- $10 cm^3$ sucrose solution
- $10 cm^3$ distilled water
- Three flat-bottomed tubes
- Clinistix reagent strips [4]
- Merckoquant nitrite-sensitive reagent sticks [1]
- Water-bath maintained at 30° C

Method

1 Introduce 25 mung beans into each of three flat-bottomed tubes. Number the tubes and make the following additions:

   Tube 1 $10 cm^3$ distilled water
   Tube 2 $10 cm^3$ sodium nitrate solution
   Tube 3 $10 cm^3$ sucrose solution

   Transfer the tubes to a water-bath maintained at 30° C.
2 Apply the following tests to the contents of each tube:

   Tube 1   Merckoquant nitrite-sensitive and Clinistix at intervals of one hour
   Tube 2   Merckoquant nitrite-sensitive reagent at intervals of four hours
   Tube 3   Clinistix reagent at intervals of one hour

   (Note that a light-coloured reading with Clinistix indicates approximately $0.25 g$ glucose/$100 cm^3$, a medium reading $0.5 g/100 cm^3$ and a dark reading $0.75 g/100 cm^3$). Record your results and draw graphs to show the induction of (i) nitrate reductase and (ii) invertase.
3 Attempt an explanation of your results.

Topics to investigate

1 Induction of enzymes in *Chlorella pyrenoidosa*. (Refer to the article by Goulding and Merrett, listed in *Further Reading*).
2 Induction of nitrate reductase in nitrogen-deficient plants.
3 Induction of maltase (β-galactosidase) in yeast grown in a solution of maltose.
4 Induction of lactase in *E. coli*.

2.19 Staining techniques for the identification of fats and proteins in maize grains

TIME 1 h

Maize grains store fats and proteins. The distribution of these compounds can be visualised by the use of biological stains, which yield coloured reaction products at room temperatures. Fats are stained blue by Sudan blue and proteins red by Ponceau S.

Preparation

Ten maize grains should be soaked in water for 48 hours. The solution of Sudan blue should be prepared by dissolving 0.2 g of the stain in absolute ethanol, while the solution of Ponceau S should contain 0.2 g dissolved in 100 cm³ water.

Materials

- Ten soaked maize grains
- 25 cm³ solution of Sudan blue[1]
- 25 cm³ solution of Ponceau S[1]
- Two 9 cm petri dishes
- Forceps
- Knife
- Blue and red pencils

Method

1 Cut each soaked maize grain in half, so that the cut bisects the endosperm and embryo.
2 Place the solution of Sudan blue into one petri dish and the solution of Ponceau S into the other. Introduce five bisected grains into each dish, ensuring that the cut surfaces are submerged.
3 After 30 minutes use the forceps to remove the stained half-grains from the petri dishes. Wash the half-grains in water and then examine them for staining. Use the coloured pencils to indicate the distribution and relative amounts of fats and proteins in the endosperm and embryo regions of the grain.

Topics to investigate

1 Distribution of fats and proteins in seeds of legumes.
2 Distribution of fats and proteins in storage organs.
3 Distribution of proteins in fruits.

Longitudinal section through the plumule of a germinating maize grain.

3 Plant Physiology, Anatomy and Morphology

3.1 Seed structure and germination in mung bean and maize

$TIME$ $1\frac{1}{2}$–2 h

This exercise involves a description and comparison of the seeds and seedlings of mung bean and maize, together with an investigation of biochemical activity within the germinating seeds.

Preparation

The bean seeds and cereal grains should be soaked in water for 24 hours before they are used in this experiment. Seedlings of the mung bean should be germinated to the stage with two foliage leaves, with the cotyledons still attached. Seedlings of maize should possess two foliage leaves.

Prepare the reagents as follows:

Stain X Dissolve 2 g iodine crystals and 4 g potassium iodide in 100 cm³ water.

Stain Y Dissolve 2.5 g silver nitrate in 100 cm³ water.

Stain Z Dissolve 0.5 g 2,3,5-triphenyltetrazolium chloride[1] in 100 cm³ water.

Materials

- Six seeds of mung bean in water
- Six cereal grains of maize in water
- Two seedlings of mung bean
- Two seedlings of maize
- 10 cm³ stain X
- 10 cm³ stain Y
- 10 cm³ stain Z
- Three petri dishes
- Forceps
- Sharp knife
- Magnifying glass (magnification ×5–10)

 WARNING: Stains Y and Z are corrosive and toxic. Avoid skin contact. Wash off any spillages with plenty of water.

Method

1 Carefully bisect each seed longitudinally, separating the cotyledons of the bean seed and cutting through both the endosperm and embryo of the maize grain.

Examine the cut surfaces of the two seeds, using a magnifying glass, and make a large, half-page labelled drawing of each.

List three major differences between the seeds.

2 The three solutions X, Y and Z are biological stains. Pour each stain into one of the petri dishes, then transfer the bisected seeds of mung bean and maize to each dish. Allow 30–40 minutes for staining, then remove the seeds and examine them. Record your observations and inferences in the form of a table.

Comment briefly on the stains X, Y and Z and on the pattern of staining shown by each seed.

3 Make scale drawings, twice life size of seedlings of mung bean and maize. Label your drawings. State three major differences between the seedlings.

Topics to investigate

1 Effects of (i) light intensity and (ii) light quality on the growth of seedlings.
2 Effects of temperature on rates of germination in seeds of different species.
3 Effects of depth of planting on (i) percentage germination and (ii) morphology of seedlings in different species.

3.2 Tests for the viability of pea and bean seeds

TIME 2 h

Two rapid biochemical tests for seed viability are introduced. You are invited to assess the relative merits and usefulness of the tests.

Preparation

Twenty soaked pea seeds and twenty soaked runner bean seeds are required. Twenty-four hours before the seeds are to be used in the experiment, each seed should be placed in a separate container, such as a flat-bottomed tube, approximately 2×8 cm. A measured quantity of distilled water should be added to each seed: 1.5 cm³ per pea seed and 2.0 cm³ per runner bean seed. The seeds should be left on the bench surface at room temperature.

A more interesting and varied result can be obtained by using old seed, preferably from the previous year's harvest. Alternatively, seed may be baked in an oven and then mixed, in different quantities, with viable seed.

Use a tetrazolium salt solution prepared by dissolving 1 g 2,3,5-triphenyltetrazolium chloride[1] in 100 cm³ water. As an economy measure, each Clinistix reagent strip may be split lengthwise into two or three pieces.

Materials

- Twenty soaked pea seeds in individual containers
- Twenty soaked runner bean seeds in individual containers
- 5 cm³ invertase concentrate[1,3,4]
- 20 cm³ tetrazolium salt solution
- Clinistix reagent strips
- Two pipette droppers
- Eight flat-bottomed tubes, each approximately 2×8 cm
- Sharp knife
- Forceps

 WARNING: Avoid skin contact with the tetrazolium salt. Wash off any spillages with plenty of water.

Method

1 Dip separate Clinistix reagent strips into the solution that surrounds several pea seeds and bean seeds. What difference do you observe? Add one drop of invertase concentrate to the solution surrounding each pea seed. Re-test each solution for glucose after a period of 30 minutes. What difference do you observe and how do you account for any change that has occurred?

2 Use the Clinistix reagent strips to test the solution surrounding each seed for the presence of glucose. A dark colour indicates approximately 0.75 g glucose/100 cm³, a medium colour approximately 0.5 g/100 cm³, and a light colour approximately 0.25 g/100 cm³.

Record your results in the form of a table and plot your results as a histogram.

3 Take one or two pea seeds from a tube that gave (i) negative, (ii) light, (iii) medium and (iv) dark readings with the Clinistix reagent strips. Use a knife to split open each seed, avoiding damage to the embryo. Transfer the split seed, or seeds, to a flat-bottomed tube and add sufficient tetrazolium salt solution to immerse the seeds. Allow the seeds to remain in the solution for 30–40 minutes, until areas of the seed have stained red. Use forceps to remove the seeds from the tetrazolium salt solution. Examine the seeds and make drawings to show the pattern of staining.

4 Attempt to correlate your results and to assess the relative merits and usefulness of each of the tests for seed viability.

Topics to investigate

1 Effects of ageing on the viability of seeds.
2 Effects of temperature of storage on the viability of seeds.
3 Use of Janus Green B (diazine green)[6] to assess the viability of seeds.
4 Use of hydrogen peroxide to assess the viability of seeds. (Viable seeds contain catalase, which catalyses the breakdown of the peroxide to oxygen and water. The level of catalase within seeds falls as ageing occurs.)

3.3 A morphological and biochemical study of normal and etiolated pea seedlings

TIME $1-1\frac{1}{2}$ h

Normal and etiolated pea seedlings are provided. This investigation involves a comparison of morphological and biochemical differences between pea seedlings grown under different conditions.

Preparation

Pea seeds should be sown in (i) light and (ii) darkness 7–10 days before they are required, and germinated at a temperature of 20–35°C. Each student will require two strips of Merck kieselgel TLC plastic sheets[1], each approximately 2.5 × 12 cm. The chromatographic solvent contains 5.5 parts cyclohexane: 4.5 parts ethyl acetate.

Materials

- Three normal pea seedlings
- Three etiolated pea seedlings
- Pestle and mortar
- Two large Kilner jars or gas jars
- Two strips of kieselgel-coated (TLC) plastic
- 5 cm³ acetone
- 20 cm³ chromatographic solvent
- Two paint brushes
- Coloured pencils (yellow, orange, yellow-green, blue-green)

Method

1 Make large, labelled drawings to show the chief differences between a normal and an etiolated pea seedling.
2 List, in the form of a table, the chief differences between the normal and etiolated seedlings.
3 Remove the leaves from three normal seedlings, transfer them to the mortar and grind them to a slurry with approximately 1 cm³ acetone. Using one of the paint brushes, paint a thin line of pigment approximately 1.5 cm from one end of one of the TLC plastic strips. Repeat the application after allowing the line to dry. Transfer the strip to one of the jars containing the chromatographic solvent, poured to a depth of approximately 1 cm. Replace the lid of the jar and allow the chromatogram to develop. Clean out the mortar, dry it and repeat the procedure with leaves from the etiolated seedling.
4 After the chromatograms have developed, use the coloured pencils to show the number and variety of pigments present in the normal seedlings. Which of these pigments have developed in the etiolated seedlings?

Topics to investigate

1 Effects of (i) light intensity and (ii) light quality on the formation of photosynthetic pigments in seedlings.
2 Effects of (i) light intensity and (ii) light quality on the morphology and anatomy of seedlings.

3.4 An investigation of the Hill reaction $\boxed{TIME \text{ 2 h}}$

The Hill reaction is named after the researcher who demonstrated that suspensions of isolated chloroplasts, extracted from cells and surrounded by an isotonic solution, would continue to perform certain reactions, as in the living cell. This investigation illustrates the ability of illuminated chloroplasts to effect reduction of two compounds that can act as hydrogen acceptors.

Preparation

The isotonic solution, in which isolated chloroplasts are suspended, is prepared by adding 10 g sucrose to 100 cm³ water. Iron(III) chloride solution, a hydrogen acceptor, is prepared by dissolving 0.25 g solute in 100 cm³ water.

Materials

- Freshly picked leaves of grass, stinging nettle or spinach
- Kitchen blender
- 100 cm³ sucrose solution
- 20 cm³ 0.01 M potassium manganate(VII) solution
- 20 cm³ 0.05 M sulphuric acid
- 10 cm³ iron(II) chloride solution
- 10 cm³ 1-volume hydrogen peroxide solution
- Merckoquant iron(II) sensitive reagent sticks[1]
- Eight flat-bottomed tubes, approximately 2 × 8 cm
- Two 100 cm³ beakers
- Two 10 cm³ plastic syringes
- Teat pipette
- Nylon stocking, muslin or cheesecloth
- Aluminium foil
- Bench lamp, fitted with a 100 W bulb

Method

1 Place approximately 25 g leaves and the sucrose solution into the blender. Grind the mixture for one minute, then filter through a nylon stocking, muslin or double layer of cheesecloth.

2 Introduce 5 cm³ of the filtrate into each of four flat-bottomed tubes, and place one tube at 5, 10, and 20 cm from the illuminated lamp. Wrap the fourth tube in aluminium foil and stand it at 10 cm from the lamp. After the tubes have stood for one hour, add to each tube: 5 cm³ sulphuric acid + 5 cm³ potassium manganate (VII) solution.

   Record the time taken for the pink–purple solution of potassium manganate(VII) to be reduced to colourless manganese(II) ions. The rate at which this occurs is directly related to the rate of hydrogen ion production by the chloroplasts.

3 Add sufficient iron(III) chloride solution to the remaining filtrate to give a concentration of 25–50 p.p.m. Fe^{2+} ions, as indicated by the iron(II) sensitive reagent sticks. Next, add the solution of hydrogen peroxide, drop by drop, until the iron(II) reagent sticks give a negative reading, following oxidation of Fe^{2+} ions to Fe^{3+} ions. Introduce 5 cm³ of the mixture into each of four flat-bottomed tubes, and set these in front of the illuminated lamp, as in procedure 2.

   After these tubes have stood for one hour, test each tube for Fe^{2+} ions, using the reagent sticks. Record your results.

4 Give an explanation of your results and state your conclusions.

Topics to investigate

1 Effects of pH on the Hill reaction.
2 Effects of (i) light intensity and (ii) light quality on the Hill reaction.

3.5 Measuring the rate of photosynthesis in a brown seaweed

TIME 2 h

Two methods are described for measuring the rate of photosynthesis in a brown seaweed, *Fucus serratus*.

Preparation

The apparatus, illustrated in Fig. 10, is constructed from a 250 cm³ medical flat, fitted with a rubber bung and a 35 cm length of 0.2 mm diameter capillary tubing. The lower end of the capillary tubing reaches almost to the base of the medical flat, while the upper end is bent, above the bung, into a horizontal position.

Materials

- Apparatus shown in Fig. 10
- Living brown seaweed (*Fucus serratus*)
- Four 100 cm³ glass beakers
- Bench lamp, fitted with a 100 W bulb
- No. 5 or 6 cork borer
- 1.5 litres sea water
- 1.5 litres distilled water
- Black paper
- Glass marking pen

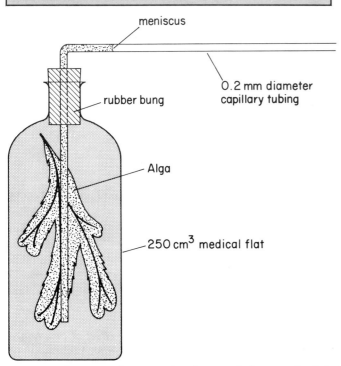

Fig. 10 *Apparatus for measuring the rate of photosynthesis in seaweed*

meniscus

rubber bung

0.2 mm diameter capillary tubing

Alga

250 cm³ medical flat

Method

1 *Method A*   Using the cork borer provided, cut 40 discs from the brown alga. Transfer ten discs to each of the beakers and add 50 cm³ sea water to each beaker. Observe that the discs, which have a density greater than that of sea water, sink to the bottom of the beakers.

  Switch on the bench lamp and position a beaker at 5, 10, 15 and 20 cm from the illuminated lamp. Record the time taken for three of the discs in each beaker either to rise to the surface, or to move from a horizontal position to 45° to the horizontal.

2 *Method B*   Take the apparatus, fill it with sea water, and replace the bung to a depth that pushes the meniscus into the horizontal part of the capillary tubing. Stand the medical flat at a distance of 5 cm from the illuminated lamp and record the distance travelled by the meniscus in a period of five minutes. Introduce a frond of *Fucus* into the apparatus. Wrap black paper around the medical flat. Position the apparatus at 5 cm from the illuminated lamp and record the distance travelled by the meniscus in a period of 5 minutes. Remove the black paper, and with the apparatus positioned at 5 cm from the lamp, record the distance travelled by the meniscus in a period of 5 minutes. Comment on your results.

3 Position the apparatus used in 2 at 5, 10, 15, 20 and 35 cm from the illuminated lamp. At each distance record the distance travelled by the meniscus in a period of 5 minutes. Record your results in the form of a table. How would you plot a graph, showing the relationship between relative rates of photosynthesis and light intensity, from these figures?

4 Explain the principles involved in each of the methods used to determine the rate of photosynthesis in *Fucus*. Which method, and why, is most satisfactory?

5 Use the apparatus illustrated in Fig. 10 to determine the effect of different dilutions of sea water on the rate of photosynthesis in *Fucus*. Describe your method and record your results as a graph.

Topics to investigate

1 Measuring the rate of photosynthesis in *Laminaria*.
2 Effects of bicarbonate ion concentrations on rates of photosynthesis in marine algae.
3 Extraction, separation and identification of photosynthetic pigments in marine algae.

3.6 The effect of bicarbonate ion concentrations on photosynthesis in Canadian pondweed

TIME 2–3 h, April–September

Bicarbonate ions are the chief carbon source for aquatic plants, such as Canadian pondweed. Using the method described, attempt to find the concentration of bicarbonate ions that is optimal for photosynthesis.

Preparation

If satisfactory results are to be obtained, it is essential to have an abundant supply of living Canadian pondweed, freshly harvested from a pond.

Materials

- Living leafy shoots of Canadian pondweed (*Elodea canadensis*)
- 20 cm³ plastic syringe
- 30 cm length of 0.1 mm diameter capillary tubing
- Rubber tubing (to fit over capillary tubing and nozzle of syringe)
- Boss and clamp
- 200 cm³ 0.1 M sodium bicarbonate solution
- 500 cm³ distilled water
- Tap water, or pond water
- 10 cm³ pipette, or plastic syringe
- Five small beakers
- Whatman narrow range pH 6–8 indicator papers[3, 4]
- Bench lamp, fitted with a 100 W bulb
- Glass marking pen

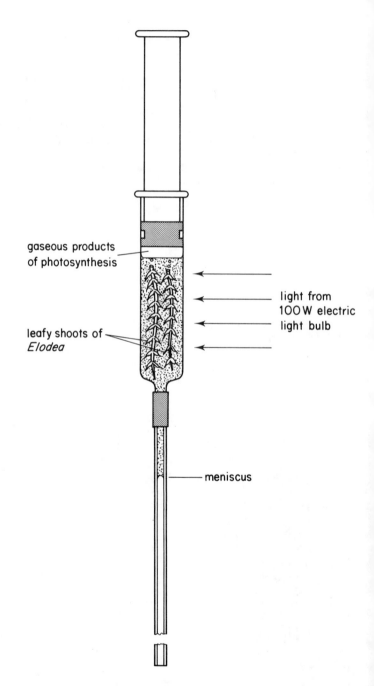

gaseous products of photosynthesis

leafy shoots of *Elodea*

light from 100 W electric light bulb

meniscus

Fig. 11 *Apparatus for determining the effect of bicarbonate ion concentration on the rate of photosynthesis*

Method

1 Cut two terminal portions of *Elodea* stem, approximately 5 cm in length. Introduce these into the barrel of the 20 cm³ syringe, with the cut ends facing away from the nozzle, as illustrated in Fig. 11. Transfer the syringe to a clamp, positioned 35–40 cm above the bench surface. Secure the syringe in a vertical position, nozzle downwards.

2 Make dilutions of the 0.1 M sodium bicarbonate solution to obtain concentrations of 0.05 M, 0.025 M, 0.0166 M, 0.0125 M and 0.01 M. Prepare 60 cm³ of each solution indicating, in the form of a table, the volume of 0.1 M sodium bicarbonate solution and distilled water used in the preparation of each solution. After preparation, measure the pH of each solution and tabulate your results.

3 Attach the capillary tube to the nozzle end of the 20 cm³ syringe by means of the rubber tubing. Make sure that the join is water-tight, and bring the end of the tubing and tip of the nozzle into close contact.

4 Placing a finger or thumb over the open end of the capillary tube, fill the barrel of the syringe to the brim with pond water or tap water. Replace the plunger of the syringe, catching any surplus water in one of the beakers. Apply gentle pressure to the plunger of the syringe until the top of the water in the barrel reaches the 20 cm³ mark. Gently raise the plunger of the syringe to draw the meniscus to a point near the top of the capillary. Mark the position of the meniscus. Set up a bench lamp, fitted with a 100 W bulb, at a distance of 10 cm from the syringe. Ensure that the shoots of *Elodea* are fully illuminated. At this point your apparatus should appear as illustrated in Fig. 11.

5 After 20 minutes measure the distance travelled by the meniscus. Record your result.

6 Repeat the procedure with 0.1 M, 0.05 M, 0.025 M, 0.0166 M, 0.0125 M and 0.01 M sodium bicarbonate solutions, allowing a little of each solution to drain through the apparatus before the syringe barrel is filled. Generally, the most satisfactory results are obtained by starting with the most dilute solution and proceeding, in order, to the most concentrated solution.
   Present your results as a graph.

7 Calculate from your results:
   (i) the concentration of sodium bicarbonate that is optimal for photosynthesis;
   (ii) the concentration of bicarbonate ions, in p.p.m. in this solution, and
   (iii) the apparent concentration of bicarbonate ions in the pond water, or tap water.
   (*Molecular weight*: $NaHCO_3 = 84$, *Atomic weights*: $C = 12$, $Na = 23$, $H = 1$, $O = 16$)

8 What conclusions do you draw regarding:
   (i) the relationship between bicarbonate ion concentrations and the rate of photosynthesis, and
   (ii) the accuracy of the results obtained?

Topics to investigate

1 Effects of light intensity on rates of photosynthesis in Canadian pondweed.
2 Effects of coloured light filters on rates of photosynthesis in Canadian pondweed.
3 Effects of pH on rates of photosynthesis in Canadian pondweed.

A cut inverted stem of Canadian pondweed in water. When the shoot is illuminated, oxygen-rich bubbles of gas are released from the leaves and the cut end of the stem.

3.7 A comparison of epidermal structure and rates of transpiration in leaves of daffodil and maize

TIME 2–3 h, February–June

Variations in the rate of transpiration from leaves of different species are determined, in part, by differences in epidermal structure. Using the materials provided, attempt to relate rates of water loss from the upper and lower surfaces of the two leaves to differences in the microscopic structure of the leaf epidermis, paying particular attention to variations in the size, form and distribution of stomata.

Preparation

Each student will require a pot-grown plant of maize, bearing not less than two foliage leaves. The leaves of daffodil should be supplied with their cut ends immersed in water.

Materials

- Leaves of daffodil, labelled A
- Pot-grown plant of maize, labelled B
- Microscope, with low-power and high-power lenses and objectives
- Four microscope slides
- Four large cover-slips
- Botanical razor, or single-edged razor blade[5]
- Ten 1 cm² pieces of cobalt chloride (or thiocyanate) paper, supplied in a petri dish
- Reel of adhesive translucent tape, 2.5 cm in width
- Scissors
- Forceps
- Ruler, graduated in millimetres and centimetres
- Fine-pointed black glass-marking pen, or overhead projector pen

Method

1 Remove one piece of dry (blue) cobalt chloride paper from the petri dish. Stick it to a piece of adhesive tape, and then fold the tape over the paper, so that the paper, sandwiched between two layers of tape, is excluded from any moisture in the air.

   Lay out five pieces of cobalt chloride paper on the bench surface, exposed to moisture in the air. At intervals of ten minutes remove one piece, enclose it within two layers of tape, and arrange the papers in a series, numbered from 0 (blue) to 5 (pink). These papers provide a standard range of colours.

2 Place one piece of cobalt chloride paper on the upper surface and a second on the lower surface of a daffodil leaf, held in place and completely covered by adhesive tape. Keep the cut end of the daffodil leaf in water. Repeat this procedure with a leaf of the potted maize plant. Observe any changes in colour that take place in the cobalt chloride papers, recording your results on a scale from 0 (blue) to 5 (pink) at intervals of 15 minutes over a period of 90 minutes. Plot your results as a graph.

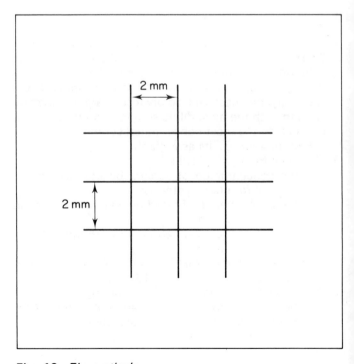

Fig. 12 *The graticule*

54

3 During this period attempt to determine the approximate number of stomata per square centimetre on both surfaces of the leaves of plants A and B. To do this, construct a graticule on the surface of one of the cover-slips by drawing three parallel lines, each 2 mm apart and three similar lines at right angles to the first three, forming four centrally positioned squares, each of area 4 mm^2, as in Fig. 12.

4 Use the graticule you have constructed, and low-powered magnification of your microscope, to estimate numbers of stomata on each surface of leaves of daffodil and maize. Record your results in the form of a table.

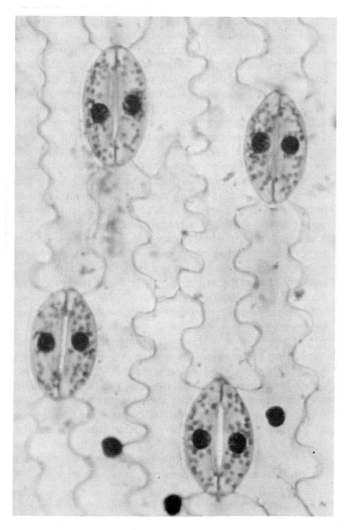

Surface view of the leaf epidermis of a monocotyledon (Lilium sp.) showing the stomata and epidermal cells (×200). In the leaf epidermis of both daffodil and maize you should be able to observe differences in the arrangement, shape and relative sizes of these cells.

5 Relate the results you obtain in part 4 to the changes recorded in part 2.

6 Do you consider that the rate of transpiration in that part of the leaf covered by paper is typical of the leaf as a whole? Give reasons for your answer.

7 In addition to the number of stomata, what structural features of the leaf epidermis could influence the rate of transpiration?

8 List six environmental factors that might influence the rate of transpiration in the potted plant.

9 Illustrate and describe the general arrangement of stomata, and the form of epidermal cells, as seen under low-power magnification of the microscope, in the lower (abaxial) epidermis of leaves from plants A and B.

10 Draw and describe a single stoma, and its surrounding cells, as seen under high-power magnification in leaves from plants A and B. Attempt to make both drawings to approximately the same scale.

Topics to investigate

1 Distribution and structure of stomata in leaves of a named plant species.
2 Loss in mass as an index of rates of transpiration in potted plants.
3 Rates of transpiration in littoral algae.

3.8 Measurement of water potential in inflorescence stalk cells of bluebell and dandelion

TIME 2½–3 h, April–June

Cut segments from the inflorescence stalks of two different plants are immersed into hypotonic, isotonic and hypertonic solutions. After observing and recording movements of the cut stalks in the different solutions, attempt to account for your results in terms of water relationships.

Preparation

The inflorescence stalks should be supplied with their cut ends immersed in water.

Materials

- Twelve stalked inflorescences of bluebell (*Endymion non-scriptus*)
- Six stalked inflorescences of dandelion (*Taraxacum officinale*)
- 250 cm³ 1M sucrose solution
- 250 cm³ distilled water
- Eight 9 cm petri dishes
- Eight 100 cm³ beakers
- Two 10 cm³, or 20 cm³, pipettes or plastic syringes
- Ruler, graduated in millimetres
- Scalpel or razor blade
- Glass marking pen

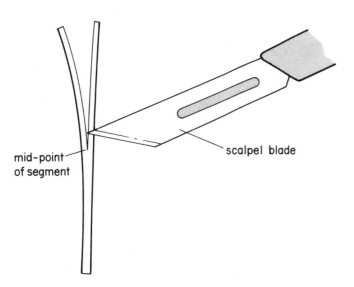

Fig. 13 *Method of bisecting an inflorescence stalk at the mid-point of a segment*

mid-point of segment

scalpel blade

Method

1 Using the molar sucrose solution and the distilled water provided, prepare 40 cm³ of each of the following: 0.1M, 0.2M, 0.4M, 0.6M and 0.8M sucrose solutions. Transfer each solution to one of the petri dishes. Label each dish. Tabulate the volumes of molar sucrose solution and distilled water used in the preparation of each solution. Construct a graph that would enable you to prepare, from molar sucrose solution and distilled water, 40 cm³ sucrose solution of any given molarity.

2 With a scalpel or sharp razor blade remove the inflorescences from their stalks; discard the inflorescences. From the inflorescence stalks of bluebell cut sixteen uniform lengths, each of 5 cm. (As far as possible use only the upper region of the inflorescence stalk, avoiding those stalks that are thick and woody.) From the inflorescence stalks of dandelion cut seven uniform lengths, each of 5 cm. Again, select material from the upper region of the inflorescence stalk.

3 Take each length of inflorescence stalk, place it on the bench surface, mark the mid-point of each length, and carefully bisect one half of the stalk longitudinally into halves of equal thickness (Fig. 13). Place two pieces of bluebell inflorescence stalk, and one of dandelion inflorescence stalk, into each of the solutions you have prepared and also into distilled water and molar sucrose. Label each dish. Leave the material for 40 minutes.

4 While you are waiting for results in 3, transfer a single cut length of bluebell inflorescence stalk to a petri dish containing distilled water. Replace the lid of the dish. Mark on the lid of the dish the mid-point between the two cut ends of the bisected half of the stalk. At intervals of five minutes, over a period of 30 minutes, measure the distance of the left-hand and right-hand halves of the stalk from the mark on the lid of the petri dish, as illustrated in Fig. 14. Tabulate your results and present them as a graph.

5 After stalks prepared in 3 have stood for 40 minutes, measure the distance between the tips of the split ends of each length of bluebell inflorescence stalk. Record your results in the form of a table, and plot your results as a graph, from which the water potential (= suction pressure) of the stalk cells could be determined. What molarity of sucrose is equivalent to the molarity of cell sap in the stalk cells?

6 Make drawings to show the appearance of lengths of bluebell and dandelion inflorescence stalks after they have stood in distilled water for 40 minutes.

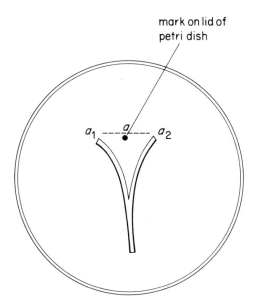

mark on lid of petri dish

a_1 - - - a - - - a_2

Fig. 14 *Method for measuring the distances of the two halves of the stalk from a mark on the lid of the petri dish. Measure the distances a–a_1 and a–a_2*

7 How do you account for the appearance of each length of inflorescence stalk after it has stood in distilled water for 40 minutes? Give your reasons.
8 (i) Suggest why strips of dandelion curve as soon as they are cut.
  (ii) Why is the curvature of cut dandelion stalks greater than that of cut bluebell stalks?
 (iii) Which plant stalk, in your view, is the most suitable for the technique employed in this investigation? Give your reasons.
9 State, with reasons, the limitations of this experiment as a method for measuring the water potential of inflorescence stalk cells.

Topics to investigate

1 Measuring the water potential of tissues by determining (i) changes in volume and (ii) changes in mass of plant material.
2 Comparative studies of the osmotic pressure of cell sap in (i) fresh water algae, (ii) estuarine algae and (iii) marine algae.

3.9 The contribution of individual leaves to the rate of transpiration in a cut shoot of beech

TIME $1\frac{1}{2}$–2h

A potometer can be used to measure the rate of water uptake by a cut leafy shoot. Using the potometer provided, carry out the following investigation to determine the contribution of individual leaves, and of the stem, to the rate of transpiration in a cut shoot.

Preparation

Each student will require a shoot of beech (*Fagus sylvatica*) bearing eight leaves.

Materials

- Leafy shoot of beech
- Simple potometer
- Boss and clamp
- Graph paper mounted on stiff card
- Bench lamp, fitted with a 60W bulb
- Top-pan balance
- Scissors
- Petroleum jelly
- Rubber tubing

Topics to investigate

1 A comparative study of water uptake and water loss by plants.
2 A comparative study of potometers.
3 Effects of anti-transpirants on rates of transpiration in a named plant.

Method

1 Set up the potometer, with the illuminated bench lamp at approximately 15 cm from the leafy twig. Number the leaves on the twig from 1 to 8, starting with the apical leaf.
2 With the leafy twig firmly secured in its position, measure the distance travelled by the meniscus in the capillary tube over a period of 5 minutes. If necessary, make a second reading. Use the formula $\pi r^2 h$ to calculate uptake of water in cm^3.
($\pi = 3.14$, r = radius of capillary in cm, h = distance travelled by meniscus in cm).
3 Apply petroleum jelly to the undersides of the two leaves closest to the stem apex (the apical pair). Measure and record the rate of water uptake. Remove the apical pair of leaves. Measure and record the rate of water uptake. Similarly, remove successive pairs of leaves from the twig, measuring and recording the rate of water uptake after removal of each pair. Retain the leaves after their removal from the stem. Finally, after removing all the leaves from the twig, measure and record the rate of water uptake by the leafless twig. Present all your results, including results from section 2, in the form of a table.
4 Plot a graph to show the effect of removing successive pairs of leaves on the rate of water uptake.
5 What percentage of the water loss occurs from the upper surface of the apical pair of leaves? Show how you arrive at your answer.
6 Comment on the result obtained after removal of all leaves from the twig.
7 Transfer the leaves to the surface of mounted graph paper. Trace around the margin of each leaf, numbering the tracings according to the position of each leaf on the twig. Cut out and weigh a $100\,cm^2$ area of mounted graph paper. Cut out and weigh each leaf tracing. Calculate leaf areas from the masses. Present your results in the form of a table.
8 Assuming that water uptake by the leafy shoot is equal to water loss, express water loss in $cm^3\,m^{-2}\,h^{-1}$ from:
  (i) the upper leaf surface;
  (ii) the lower leaf surface;
  (iii) the upper and lower leaf surface.
  Show how you arrive at your answers

3.10 A volumetric method for investigating the water relations of succulent tissues

If large rod-like pieces of succulent tissues, such as the tubers of potatoes or roots of swedes, are placed into a measured volume of either a hypotonic solution, or a hypertonic solution, there is a change in the volume of the external solution, as water enters the tissue or is drawn from it. Only when a tissue is surrounded by an isotonic solution is there no change in the volume of the external solution. This exercise makes use of that principle in an investigation of water relationships.

Preparation

$180 \, cm^3$ of each of the following will be required:
 (i) tap water
 (ii) 0.25 M sodium chloride solution
 (iii) 0.5 M sodium chloride solution
 (iv) 0.8 M sodium chloride solution
 (v) 1 M sodium chloride solution
 (vi) Solution X (sodium chloride solution, prepared by the supervisor, its molarity not revealed to the student).
Twenty-four hours before the experiment, pour $90 \, cm^3$ of tap water, or $90 \, cm^3$ of each saline solution, into each of six containers, such as beakers or paper cups. Use a no. 13 cork borer (diameter 1.5 cm) to cut solid rods of tissue from potato tubers. Trim each rod to 3 cm in length. Transfer four rods to each container with tap water or saline solution. Ensure that the rods are submerged. Similarly, set up six beakers or paper cups containing an identical set of solutions. Place four rods of swede root tissue into each beaker or cup. Finally, set up two containers each containing $90 \, cm^3$ of solution X. Add four rods of potato tissue to one container and four rods of swede tissue to the other. Allow all the containers, labelled with their contents, to stand for 24 hours, uncovered, on the bench surface.

Materials

- Rods of potato tuber, immersed in tap water or saline solution
- Rods of swede root, immersed in tap water or saline solution
- Rods of potato and swede, immersed in solution X
- $100 \, cm^3$ measuring cylinder
- Iodine solution (for starch test)
- Clinistix reagent strips and invertase concentrate (for sucrose test)

Method

You are provided with six different samples of potato tissue, and six different samples of swede root, in separate containers. Each sample has stood for 24 hours in $90 \, cm^3$ of one of the following, as indicated on the container:

 (i) tap water
 (ii) 0.25 M sodium chloride solution
 (iii) 0.5 M sodium chloride solution
 (iv) 0.8 M sodium chloride solution
 (v) 1 M sodium chloride solution
 (vi) solution X (a saline solution of unspecified molarity).

1 Using the $100 \, cm^3$ measuring cylinder provided, measure the volume of tap water and of the saline solutions, including solution X, that now surrounds each sample of tissue. Present your results logically in a table.
   Plot a graph of your results in a way that will enable you to determine both the molarity of solution X and the osmotic potential (= osmotic pressure) of the potato and swede saps. What is the molarity of solution X?
2 If a molar saline solution exerts an osmotic pressure of 22.4 atmospheres, calculate, in atmospheres, the osmotic potential of the potato sap and the swede sap.
3 Suggest three improvements in technique that would have enabled more accurate results to be obtained.
4 Use the reagents provided to test the tissues for starch and sucrose. In the test for sucrose, crush a little of the tissue, add one drop of invertase concentrate to it, wait for 3–5 minutes, then apply a Clinistix reagent strip. A positive result is indicated by a colour change from pink to blue in the Clinistix. How could differences in the nature of the stored carbohydrates in potato and swede account for differences in the osmotic potential of their cell saps?

Topics to investigate

1 An investigation of the water relations of tissues from root crops.
2 An investigation of the water relations of tissues from succulent fruits.

3.11 To determine the conditions necessary for germination of pollen grains of Amaryllis

TIME 1–2 h, December–April

Pollen grains from different plants have different requirements for their germination. Whilst some will germinate in water only, others require to be immersed in solutions of glucose or sucrose, often within a narrow range of concentration. You are provided with pollen grains of the lily *Amaryllis hippeastrum*, which have stood for 24 hours in distilled water, solutions of glucose or solutions of sucrose. Carry out the following exercise in an attempt to determine the conditions required for germination of the pollen grains of this species.

Preparation

Pollen should be removed from flowers of *Amaryllis* 2–3 days after the flowers have opened. Samples of pollen should stand for 24 hours in the following:
  (i) distilled water
 (ii) glucose solutions containing, respectively, 3, 6, 9, 12, 15, 18, 21 and 24 g glucose/100 cm^3
(iii) sucrose solutions containing, respectively, 3, 6, 9, 12, 15, 18, 21 and 24 g glucose/100 cm^3.
These preparations should be maintained at a temperature within the range 20–25° C in order to encourage germination.

Results obtained by placing mixtures of pollen and sugar solutions on a slide, covered by a cover-slip, are rarely satisfactory as the water tends to evaporate, causing the concentration of the sugar solution to increase. A simple piece of apparatus can be made using a slide, a ring of plastic cut from tubing or from a plastic syringe, and a cover-slip, as illustrated in Fig. 15. The ring of plastic is placed on the slide and cemented into position by application of a translucent glue, or perspex cement, on the outside. After allowing time for the glue to dry, a paint brush is used to transfer pollen to the centre of the ring. A single drop of distilled water, or solution of sugar, is then added. After greasing the rim of the plastic ring with petroleum jelly, a cover-slip is put into place over the ring to prevent loss of water by evaporation.

Materials

- Pollen grains of *Amaryllis*, mounted in distilled water
- Pollen grains of *Amaryllis*, mounted in glucose solutions (3, 6, 9, 12, 15, 18, 21 and 24 g glucose/100 cm^3)
- Pollen grains of *Amaryllis*, mounted in sucrose solutions (3, 6, 9, 12, 15, 18, 21 and 24 g sucrose/100 cm^3)
- Compound microscope, fitted with low-powered lenses and objectives
- Large cover-slip
- Fine-pointed glass-marking pen, or overhead projector pen

Method

1 As in exercise 3.7, construct a graticule by marking squares of area 2 mm^2, or 4 mm^2, on the surface of the cover-slip. Examine each preparation under the microscope. Use the graticule to determine the percentage germination of pollen grains in: (i) distilled water, (ii) glucose solutions and (iii) sucrose solutions. Record your results in the form of a table. Plot your results as a graph.
2 Make labelled, scale drawings to show the appearance and structure of pollen grains in each solution. Present your drawings in a clear, logical sequence.
3 How do you account for the results obtained in parts 1 and 2 of this investigation?

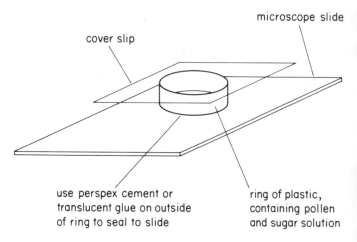

cover slip

microscope slide

use perspex cement or translucent glue on outside of ring to seal to slide

ring of plastic, containing pollen and sugar solution

Fig. 15 *Apparatus for investigating the effect of different solutions on the germination and growth of pollen grains*

Topics to investigate

1 Factors influencing the germination of pollen grains from different plant species.

3.12 | A photometric method for measuring the growth rate of populations of yeast cells

TIME 2–3 days

An accurate photometric method is described for measuring increases in the population density of yeasts and other microorganisms. Carry out the exercise to determine the rate of population growth of yeast in a solution of glucose.

Preparation

Yeasts are generally available in dried form from shops which stock materials for home-brewing and wine-making.

Materials

- Dired yeast (*Saccharomyces cerevisiae*)
- Light meter[2]
- Electric lamp, fitted with a 40 W bulb
- 200 cm³ 0.5 M glucose solution
- 1 g ammonium phosphate
- 25 cm³ pipette
- Ten 250 cm³ conical flasks
- Ten 9 cm petri dishes
- Boss and clamp
- Water-bath maintained at 35° C
- Glass marking pen

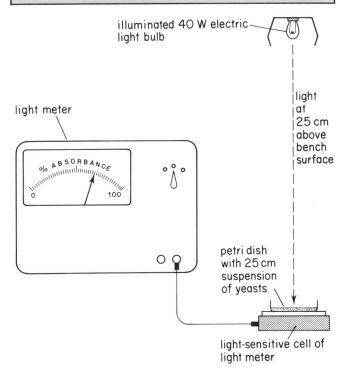

Fig. 16 *Apparatus for measuring the growth rate of a population of yeast cells*

Method

1 Weigh out samples of yeast in each of the following quantities: 1.0, 1.25, 1.5, 1.75, 2.0, 2.25, 2.5, 2.75 and 3.0g. Transfer each sample to a labelled conical flask and add 100 cm³ water to each flask. Allow the mixture to stand for approximately 30 minutes, stirring at intervals to ensure even distribution of the yeast cells.

2 Pipette 25 cm³ of the suspension containing 1 g yeast to a petri dish. Place the dish over the light-sensitive cell of the light meter and, with the 40 W electric light blub illuminated, and clamped, at 25 cm above the bench surface, as in Fig. 16, read off the percentage absorbance from the scale.

   Repeat this procedure for each suspension of yeast. Record your results in the form of a table. Plot your results as a graph.

3 Introduce 2 g dried yeast into a conical flask and add 200 cm³ 0.5 M glucose solution together with 1 g ammonium phosphate. Transfer the conical flask to a water-bath maintained at 35° C. At intervals of 6 hours over a period of 48 hours withdraw 25 cm³ samples and measure the density of the population of yeast cells. From your standard curve, read off the dry mass of the yeast. Record your results in the form of a table. Plot your results as a graph.

Topics to investigate

1 The growth rate of populations of yeast cells in solutions of glucose, sucrose and maltose.
2 Effects of (i) ammonium phosphate and (ii) yeast extract on the growth rate of yeast populations.
3 Effects of ethanol on the growth rate of populations of yeast cells.
4 Effects of (i) temperature and (ii) pH on the growth rate of populations of yeast cells.
5 Growth rates of bacterial populations.
6 Growth rates of populations of isolated animal cells or plant cells.

Fermentation of sugars by yeast (*Saccharomyces cerevisiae*) is accompanied by the evolution of carbon dioxide. Measurements of the volume of gas evolved, as in this exercise, provide an index of the rate at which fermentation is taking place.

Preparation

Prepare the suspension of yeast by adding 10 g dried yeast to 100 cm³ water. Activate the yeast by adding a level teaspoon of glucose or sucrose, mix thoroughly, and allow the mixture to stand overnight in an open beaker or similar container.

Materials

- 40 cm³ activated yeast suspension
- 40 cm³ 0.25 M glucose solution
- 40 cm³ 0.25 M sucrose solution
- 40 cm³ 0.25 M maltose solution
- 40 cm³ 0.25 M lactose solution
- Four 10 cm³ plastic syringes
- 10 cm³ pipette or plastic syringe
- Four bosses and clamps
- Four 100 cm³ beakers
- Four 30 cm lengths of 1.0 mm diameter capillary tubing
- Rubber tubing (to fit over capillary tubing and nozzle of syringe)
- Ruler, graduated in millimetres
- Glass-marking pen

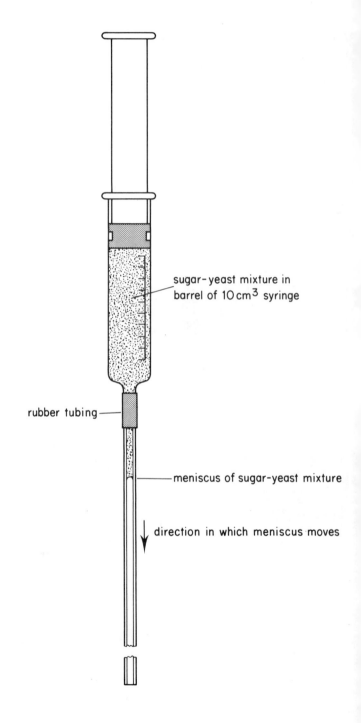

sugar-yeast mixture in barrel of 10 cm³ syringe

rubber tubing

meniscus of sugar-yeast mixture

direction in which meniscus moves

Fig. 17 *Apparatus for measuring the rate of fermentation of sugar by yeast*

Method

1 Thoroughly mix the suspension of yeast cells by shaking. Pipette 10 cm³ of this suspension into each of four beakers. Number the beakers 1 to 4. Introduce 20 cm³ glucose solution into beaker 1. Repeat this procedure with solutions of sucrose, maltose and lactose, added to yeast in each of beakers 2, 3 and 4.

2 Introduce 10 cm³ of the glucose–yeast mixture into one of the 10 cm³ plastic syringes, drawing in the mixture via the nozzle. Fill each of the three remaining syringes with one of the yeast–sugar mixtures contained in beakers 2, 3 and 4.

3 Transfer each filled syringe to a clamp, at a height of 35–40 cm above the bench surface. Secure the syringe into position, nozzle pointed downwards. Attach a 2–3 cm length of rubber tubing to one end of the capillary tube. Attach the free end of the rubber tubing to the nozzle of the syringe; bring the end of the tubing and the tip of the nozzle into close contact, as illustrated in Fig. 17.

4 Apply gentle pressure to the plunger of the syringe until a meniscus appears at the top of the tube. Mark the position of this meniscus with a glass-marking pen. As carbon dioxide gas is evolved during fermentation of the sugar, each bubble of gas will displace an equivalent volume of the mixture into the capillary tube. At intervals of 15 minutes, over a period of 90 minutes, record the length of the column of mixture in each capillary tube. Record your results in the form of a table. Plot your results as a graph.

5 Attempt an explanation of your results, giving a full explanation of the principles on which the technique is based, and a criticism of the methods employed.

Topics to investigate

1 Rates of fermentation of glucose by different species of yeast.

2 Rates of fermentation of yeast in solutions of glucose, maltose and sucrose.

3 Comparative studies of rates of fermentation by different strains of *Saccharomyces cerevisiae*.

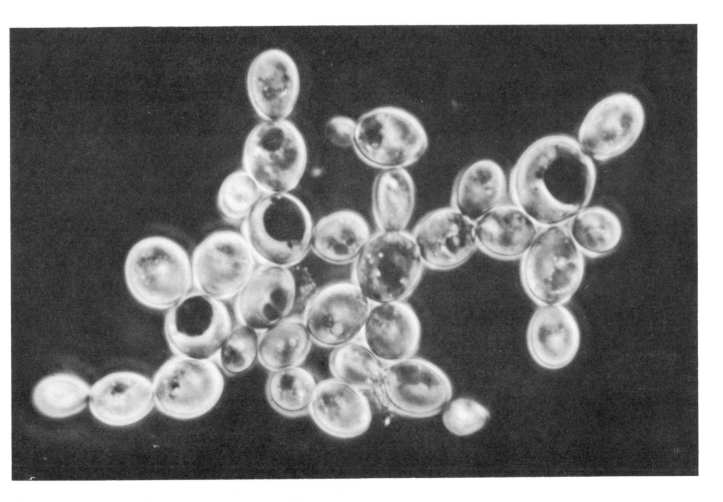

A colony of yeast cells (Saccharomyces cervisae) ×225.

3.14 The effect of ethanol on the rate of anaerobic respiration of glucose by yeast

TIME 2 h

Rates of respiration in living organisms are often expressed in terms of a respiratory quotient, Q_{10}.

$$Q_{10} = \frac{\text{rate of the reaction at } t + 10^\circ C}{\text{rate of reaction at } t^\circ C}$$

Rates of respiration in this exercise are determined by a colorimetric method, involving a colour change from purple–red to colourless in alkaline phenolphthalein indicator, a pH indicator, as a result of increased acidity resulting from the fermentation of glucose.

Preparation

Alkaline phenolphthalein indicator is prepared by the addition of 5 cm³ bench solution of phenolphthalein to 100 cm³ 0.1 M sodium hydroxide. The mixture should be prepared immediately before it is required. The suspension of yeast cells is prepared by adding 10 g dried yeast to 100 cm³ water.

Materials

- Suspension of yeast cells (100 cm³)
- 100 cm³ 0.5 M glucose solution
- 25 cm³ absolute ethanol
- 100 cm³ alkaline phenolphthalein indicator
- 100 cm³ distilled water
- Eighteen flat-bottomed tubes
- Water-baths maintained at 20, 30, 40 and 50°C
- Two 5 cm³ pipettes or 5 cm³ plastic syringes

Method

1 Introduce the following into each of eight flat-bottomed tubes:
   (i) 5 cm³ suspension of yeast cells
   (ii) 5 cm³ glucose solution
   (iii) 5 cm³ alkaline phenolphthalein indicator
   (iv) 5 cm³ distilled water.
   Transfer two of the tubes to each of the four water-baths. Remove the tubes as soon as the mixture is colourless. Record, in the form of a table, the times taken for decolourisation to occur at each temperature. Plot your results as a graph and calculate Q_{10} values. Comment briefly on your results.

2 Introduce the following into each of two flat-bottomed tubes:
   (i) 5 cm³ suspension of yeast cells
   (ii) 5 cm³ glucose solution
   (iii) 5 cm³ alkaline phenolphthalein indicator
   (iv) 5 cm³ absolute ethanol.
   This mixture contains 25% ethanol by volume. Transfer the tubes to a water-bath maintained at 30°C.

   Prepare, in duplicate, mixtures containing, respectively, 12.5, 6.25, 3.125 and 1.5625% ethanol by volume. Transfer these tubes to a water-bath maintained at 30°C, and record the time taken for each mixture to decolourise. Record your results in the form of a table. Plot your results as a graph, and comment briefly on them.

Topics to investigate

1 Effects of reducing agents and oxidising agents on the rate of respiration in yeast.
2 Effects of different alcohols on the rate of respiration in yeast.
3 Effects of substrate concentration on rates of respiration in yeast.

3.15 | Arrangement and structure of vascular elements in petioles of celery

TIME $2\frac{1}{2}$–3h, July–November

This is an investigation into the arrangement and structure of vascular elements in leaf petioles of celery.

Preparation

The red dye for staining the walls of xylem vessels contains 0.1 g eosin[6] + 0.1 g rose bengale[6] in 100 cm³ water. Prepare the iodine solution by dissolving 2 g iodine crystals and 4 g potassium iodide solution in 100 cm³ water.

Materials

- Celery plant, with intact leaves
- 50 cm³ red dye
- Iodine solution
- Dissecting dish, containing black wax, covered by water
- 500 cm³ beaker
- Microscope slides and cover-slips
- Compound microscope, fitted with low-powered and high-powered lenses and objectives
- Bench lamp, fitted with a 60 W bulb
- Two sewing needles
- Pins
- Ruler, graduated in millimetres
- Scalpel

Method

1 Introduce 50 cm³ of the red dye into the beaker. Remove 3–4 outer leaf petioles from the celery plant and transfer these to the beaker, with the base of each petiole standing immersed in the dye. Place the illuminated bench lamp behind one of the petioles and observe the dye as it ascends through the vascular bundles. Make any measurements that you consider necessary and write a concise description of your observations.

2 Remove part of a vascular bundle, approximately 1 cm in length, from one of the petioles occupying the mid-region of the stem. Transfer this material to a microscope slide and add two drops of iodine solution. Use the two needles to tease the vascular tissue, until elements of the xylem and phloem have been separated. Cover the material with a cover-slip, transfer it to the stage of your microscope, and make labelled drawings to show the appearance of conducting elements from the xylem and the phloem.

3 After the dye has penetrated to the leaflets, transfer one of the leaves to a dissecting dish, hold it in position with a pin at each end and cut off the leaflets at a distance of 4–8 mm from the petiole. Make a careful dissection of the petiole to display the vascular bundles, using the scalpel to remove all overlying tissues.

4 By means of an annotated drawing, illustrate the arrangement of vascular bundles in the petiole. You are not, however, expected to show the origin of bundles which supply the leaflets.

Topics to investigate

1 Arrangement of the vascular bundles in stems of broad bean, white deadnettle or cow parsley.
2 Leaf venation in dicotyledons.
3 Leaf venation in monocotyledons.
4 Petiolar venation in dicotyledons.

3.16 Isolation of antheridia and archegonia from moss plants

TIME 30 *min, March–July*

The sex organs of moss plants, antheridia and archegonia, are borne at the apex of plants, surrounded by a terminal whorl of leaves. Removal of intact antheridia and archegonia, which are visible only with the aid of a microscope, requires patient and careful dissection.

Preparation

Satisfactory results can generally be obtained with plants of the genera *Mnium*, *Funaria* and *Polytrichum*, freshly gathered, and presented to students in a sealed polythene bag, or other air-tight container.

Materials

- Clump of moss, containing male and female plants
- Compound microscope, fitted with a low-powered lens and objective
- Two microscope slides and cover-slips
- Two sewing needles

Method

1 Using the needles provided, and the microscope, isolate a single mature antheridium and a single mature archegonium from the moss plants. Mount the antheridium in water on a slide and make an accurate drawing to show how it appears when viewed beneath the low-power magnification of your microscope. Similarly, mount and draw a mature archegonium.

Topics to investigate

1 Isolation of embryos from the ovules of shepherd's purse.

Moss plants × 7. Archegonia and antheridia are surrounded by the terminal whorl of leaves.

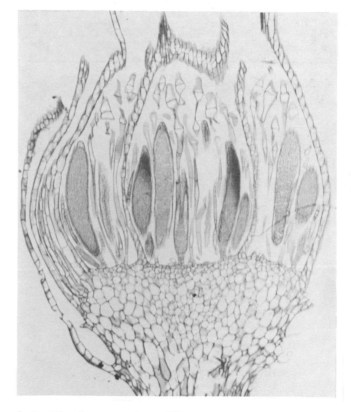

Antheridia of a moss plant × 20.

Variety of structure in epidermal trichomes

Trichomes are unicellular or multicellular hairs associated with the epidermis of vascular plants. Attempt to make accurate, unlabelled drawings of trichomes from the three plants provided.

Preparation

Many plants are suitable for this exercise. The nature of the material presented to students will obviously depend on the availability of species.

Materials

- Leafy shoot of stinging nettle (*Urtica dioica*)
- Leafy shoot of bramble (*Rubus fruticosus*)
- Flowers of herb robert (*Geranium robertianum*)
- Compound microscope, fitted with low-powered and high-powered lenses and objectives
- Three microscope slides and cover-slips
- Scalpel

Method

1 Use the scalpel to strip off a piece of epidermis from the leaf petiole of stinging nettle, mount it in water on a slide, transfer it to the stage of the microscope, and make an accurate unlabelled drawing to show the appearance of all types of trichome that are present.
2 Strip a piece of epidermis from the apical region of bramble. Mount the strip in water, examine it beneath the microscope, and make accurate drawings of all types of trichome that are present.
3 Remove a sepal from the flower of herb robert, mount it and view it beneath the microscope. Make accurate, unlabelled drawings of all types of trichome that are present.

Topics to investigate

1 Structural variations of epidermal trichomes in a named family of angiosperms.
2 Structural variations of epidermal trichomes in fern sporophytes.

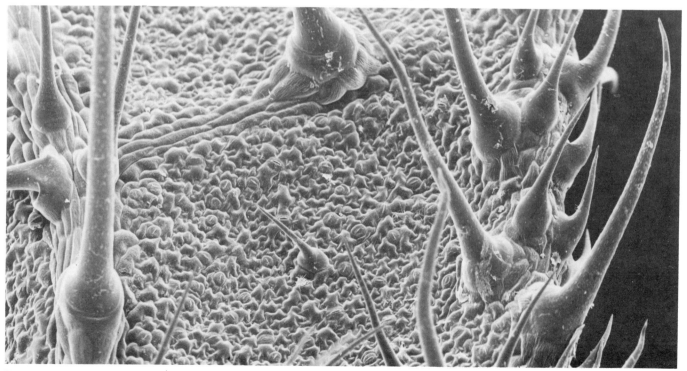

Epidermal trichomes of Urtica dioica ×200.

3.18 Morphology and cellular anatomy of four filamentous green algae

TIME 2 h

This exercise involves a morphological and anatomical study of four filamentous green algae, which exhibit variation in the pattern of their branching and the fine structure of their filaments.

Preparation

The algae required for this exercise, available from several suppliers of botanical material, will remain viable for several months if they are kept in small individual containers, shaded from high light intensity, and maintained within the temperature range 5–20° C.

Materials

- Living culture of *Spirogyra varians*,[4] labelled A
- Living culture of *Zygnema* sp.,[4] labelled B
- Living culture of *Cladophora crispata*,[4] labelled C
- Living culture of *Hydrodictyon*,[4] labelled D
- Iodine solution (for starch-testing)
- Four microscope slides and cover-slips
- Compound microscope, fitted with low-powered and high-powered lenses and objectives
- Forceps
- Mounted needle
- Glass-marking pen

Method

1 Using the low-powered lens and objective of your microscope, make outline diagrams to show the form of branching, or arrangement of cells, in each of the four algae A, B, C and D. (Detailed drawings of cells are not required.)
2 Using the high-powered magnification of your microscope, make large labelled drawings to show the appearance of one cell from each specimen of A, B and C. Label the following structures in your drawing: cell wall, chloroplast, pyrenoid, nucleus, starch grain, vacuole.
3 Make a table to show the main similarities and differences between specimens A, B, C and D.

Topics to investigate

1 Cellular anatomy of four unicellular green algae.
2 Morphology and cellular anatomy of four brown, or red, algae.

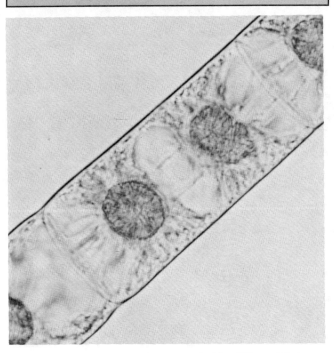 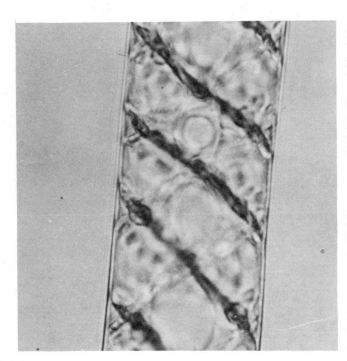

Living vegetative cells of Zygnema (left) ×150 and Spirogyra (right) ×250.

4 Animal Physiology, Behaviour and Structure

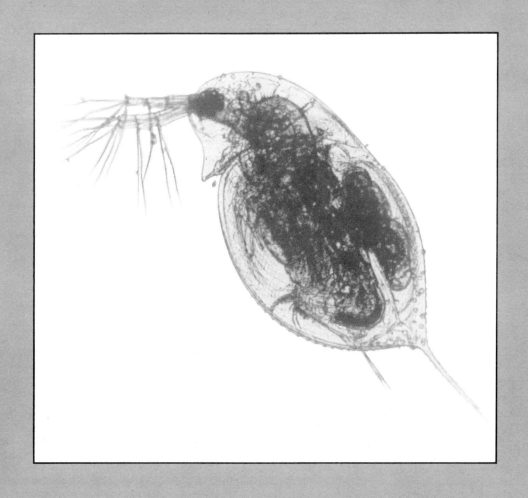

Active transport of chloride ions across the skin of a frog

TIME 1½–2h

The skin of an amphibian, such as a frog, acts as a semi-permeable membrane, but may retain sodium chloride within the body of the animal by means of an active mechanism. This is an investigation of the rate at which chloride ions are transported across the skin of a freshly-killed frog, from the outside to the inside, and from the inside to the outside.

Preparation

The frog should be killed immediately before it is required, using a killing agent such as ethyl acetate or chloroform.

Materials

- Freshly-killed frog
- Dissecting dish, containing black wax covered by distilled water
- Dissecting instruments
- Pins
- Two flat-bottomed tubes, approximately 2 × 8 cm
- Quantab chloride titrator no. 1175
- 1 cm³ plastic syringe
- 10 cm³ pipette, or plastic syringe
- 5 cm³ 0.5 M sodium chloride solution
- Cotton

Method

1 Using scissors and a scalpel, remove the skin from the thigh and shank region of the hind limb. Firstly, excise the foot at the ankle joint, then release the skin from underlying connective tissue around the top of the leg, as illustrated in Fig. 18. By gently pulling the skin from around the top of the thigh, peel off the skin from the leg, releasing it at the knee from underlying tissue. Do not puncture the skin in process of its removal.
2 Similarly, remove the skin from the other hind limb.
3 Using forceps, invert one of the pieces of skin so that the skin surface is on the outside. You should now have two similar pieces of skin, one of which is inside-out.
4 Close the ankle-end of each piece of skin by tying it with cotton. Obtain the help of an assistant to introduce 1 cm³ 0.5 M sodium chloride solution into each sac of skin, then tie the skin at the top to prevent leakage of the saline solution. Wash the outsides of each sac of skin in distilled water.
5 Introduce 5 cm³ distilled water into each of two flat-bottomed tubes. Transfer the inside-out sac of skin to one tube, and the outside-out sac of skin to the other. At intervals of 15 minutes, over a period of 45 minutes, record the concentration of chloride ions in the external solution, using the chloride titrator. Record your results in the form of a table. Construct a graph from your results.
6 How do you account for your results?

Topics to investigate

1 Use of reagent sticks to determine rates of Na⁺ and K⁺ transport across the skin of a frog.
2 Transport of anions across the skin of a frog.
3 Transport of glucose across the skin of a frog.

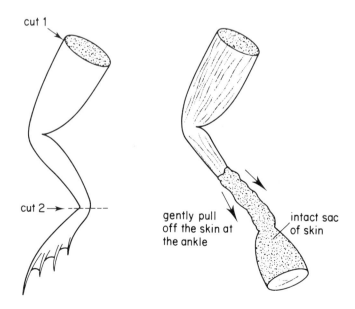

cut 1

cut 2

gently pull off the skin at the ankle

intact sac of skin

Fig. 18 *Removal of skin from the hind leg of a frog*

4.2 An investigation of the action of enzymes in the alimentary canal of a freshly-killed mammal

Using a freshly-killed mammal, carry out the following exercise to identify those regions of the alimentary canal that contain active carbohydrases, lipases and proteolytic enzymes.

Preparation

The agar plates, containing enzyme substrates, should be prepared from agar powders of high quality, free from ash, such as no. 1, no. 3 or bacteriological agar from the Oxoid range. Each agar plate, containing the materials listed below, should be prepared by pouring molten agar into a 9 cm petri dish, to cover the floor of the dish to a depth of 2–3 mm only.

Starch agar 100 cm³ distilled water, 1 g soluble starch, 2.5 g agar powder.
Mix the starch with a little cold water and add this mixture, drop by drop, to boiling water, stirring to ensure even distribution.

Milk agar 100 cm³ distilled water, 2 g Marvel milk, 2.5 g agar powder.
Mix the milk powder with a little cold water and add this, drop by drop, to boiling water, stirring to ensure even distribution.

Salad cream agar 100 cm³ water, two level teaspoons of salad cream, 0.5 g sodium carbonate, 2.5 g agar powder.
Mix the salad cream and sodium carbonate with cold water and gently heat the mixture until it boils, stirring to ensure even distribution.

The solutions of sucrose, maltose and lactose are each prepared by adding 10 g of each sugar to 100 cm³ water. Use a solution of iodine suitable for starch testing. The potassium manganate(VII) solution contains 2 g of the solute dissolved in 100 cm³ water.

Materials

- Freshly-killed mammal
- Dissecting instruments, board, awls etc.
- Six small beakers, covered with aluminium foil
- pH papers (pH 3.0–9.0 range)
- Two starch agar plates
- Two milk agar plates
- Two salad cream agar plates
- Solution of iodine
- 10 cm³ potassium manganate(VII) solution
- 20 cm³ sucrose solution
- 20 cm³ maltose solution
- 20 cm³ lactose solution
- 18 flat-bottomed tubes, each approximately 2 ×8 cm;
- 10 cm³ pipette, or plastic syringe;
- Clinistix reagent strips
- Incubator or oven maintained at 35°C
- Glass-marking pen

Method

1 Dissect the mammal provided to expose the alimentary canal. Remove the following regions only.
   - (i) Oesophagus
   - (ii) Stomach.
   - (iii) 10 cm length of the duodenum, cut posteriorly to the stomach.
   - (iv) 10 cm length from the mid-small intestine.
   - (v) 10 cm length of ileum, cut immediately anterior to the caecum.
   - (vi) Colon.

   Transfer each region of the alimentary canal to a small beaker, covering it with aluminium foil to prevent desiccation. Label the beakers. Measure the pH of each region and record your results in the form of a table.

2 Cut from each region of the alimentary canal an area of approximately 1 cm². Transfer this material to the starch agar plate, positioned so that the mucosal lining is in contact with the substrate. Label the position of each region of the gut on the lid of the dish. Similarly, spread the surface of the milk agar and salad cream agar with 1 cm² pieces cut from each region of the alimentary canal. Set up a second dish of each type.

3 Allow the agars to stand for 1 hour in an oven, or incubator, maintained at 35°C. After removing the agars:
   - (i) overpour the starch agar with iodine solution, and
   - (ii) overpour the salad-cream agar with potassium manganate(VII) solution.

   The milk agar will require no further treatment.

   Wash off the iodine solution after 2–3 minutes, and the potassium manganate (VII) after 5–8 minutes. If digestion of substrate has occurred, a translucent area will have developed beneath and around the tissue. The total area of the translucent region formed bears a direct relationship to the amount of enzyme present.

4 Transfer 2 cm³ sucrose solution to each of six tubes. Introduce into each tube a piece of tissue taken from each region of the alimentary canal, either 1 cm² or a 0.5 cm length. Allow the tube to stand in the oven or incubator for 1 hour at 35°C, then test the sugar solution for the presence of glucose, using a Clinistix reagent strip. A positive test indicates the presence of invertase:

$$\text{Sucrose} \xrightarrow{\text{Invertase}} \text{Glucose} + \text{Fructose}$$

5 Repeat procedure 4 with solutions of maltose and lactose. A positive test with maltose indicates the presence of maltase.

$$Maltose \xrightarrow{\text{Maltase}} Glucose$$

A positive test with lactose indicates the presence of lactase.

$$Lactose \xrightarrow{\text{Lactase}} Glucose + Galactose$$

6 Record all your results in the form of a table. Comment on the distribution of carbohydrases, lipases and proteolytic enzymes, within different regions of the alimentary canal.

Topics to investigate

1 Enzyme activity in (i) tissues and (ii) organ systems of a mammal. (Many tissues and organs of a mammal, in addition to those of the alimentary canal, contain active enzymes, especially amylase.)
2 Identification of enzymes in soil samples. (Soils contain active carbohydrases, lipases and proteolytic enzymes on account of populations of micro-organisms that are present.)
3 Enzyme activity in germinating seeds.

Photomicrograph of a stained section of the stomach wall of a rat. Secretory cells form the upper section.

4.3 An investigation of human sensory receptors

TIME 1h

In this investigation you are invited to survey the distribution, functions and limitations of some of the sense organs in your body.

Preparation

Two pins should be mounted in a cork as illustrated in Fig. 19.

Materials

- Cork with two mounted pins
- Half a postcard (cut lengthwise)
- Small pocket mirror
- Dry cotton wool
- Five cotton buds
- Black or blue ball-point pen
- 1 cm³ dilute sodium chloride solution
- 1 cm³ dilute solution of sucrose
- 1 cm³ lemon juice
- 1 cm³ quinine
- 1 cm³ water.

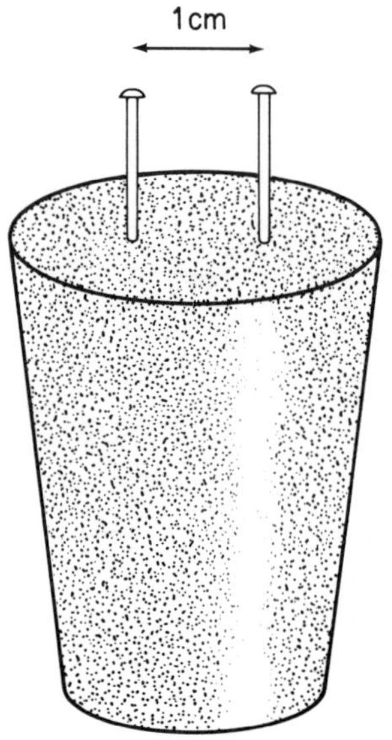

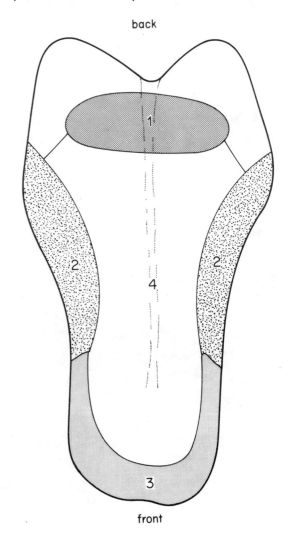

Fig. 19 *Arrangement of pins in the cork*

Method

1 Take the dry cotton wool and stroke it lightly over (i) the palm of the hand and (ii) the forearm. Describe the sensations experienced and attempt to account for them.

2 Rub the palm of the right hand vigorously backwards and forwards for a total of 15–20 movements against: (i) the palm of the left hand; (ii) the back of the left hand and (iii) the left forearm. Describe the sensations experienced and attempt to account for them.

Fig. 20 *Regions of the human tongue*

3 Take the cork with two mounted pins, close your eyes, and apply the pin heads to the following regions in turn: (i) tip of thumb, (ii) base of thumb, (iii) back of hand; (iv) forearm; (v) lips and (vi) forehead. In which of these regions are you able to detect two distinct points of contact? What explanation can you offer?

4 Take the ball point pen in your right hand, between the thumb and index finger. Strike one end of the pen, with a sharp beating action, against: (i) the palm of the left hand, (ii) the back of the left hand, (iii) the forearm, (iv) the side of the face. Attempt to make each stroke of equal intensity. List these regions in order of increasing pain experienced.

5 Dip a cotton bud into each of the following: (i) solution of sodium chloride (salt), (i) solution of sucrose (sweet), (iii) lemon juice (sour), and (iv) quinine (bitter). Apply each compound, in turn, to each of the four regions of the tongue, illustrated in Fig. 20. List the regions of the tongue in which each compound is tasted.

Dip a fifth cotton bud into water. Attempt to determine if a particular region of the tongue is sensitive to water.

your right hand. Close your left eye. Look at the left-hand circle with the right eye, and move the card backwards and forwards until the right-hand circle disappears from view. Measure and record the distance between the left-hand circle and the right eye.

Repeat the exercise with your left eye and record your results.
   (i) Why is it that, at a particular distance, one of the circles disappears from view?
   (ii) What name is given to this distance?
   (iii) If on some subsequent occasion you repeated this test and found that the distance had lengthened, what conclusion would you reach? What would be the consequence of this change and how might it be corrected?

7 Close the left eye. Focus on the corner of the room, where two walls and the ceiling meet. Hold the pen at about 20 cm from the eye, in line with the corner of the room. Attempt to bring both the pen and the corner of the room into sharp focus. Describe the nature of the visual problem and attempt an explanation.

8 If you were provided with a clock in which the ticking action could be heard by most people at a distance of from 1–1.5 m, describe how you would attempt to compare the relative efficiency of hearing in the left and right ears of three subjects.

Topics to investigate

1 Use of an audiometer to measure the frequency range of the human ear.
2 Variations in eyesight with age and sex.
3 Optical illusions.

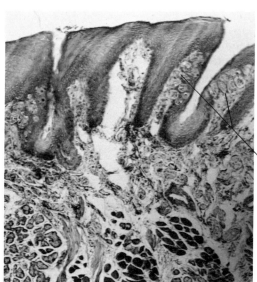

Taste buds

A vertical section of the surface of the tongue.

6 Use the ball-point pen to draw two circles, each approximately 1.5 cm in diameter, one at each end of the half postcard, approximately 0.5 cm from the edge. Fill in each circle with ink. Take up a position with the right eye at the same level as the bench surface, positioned at the edge of the bench, looking along the surface. Hold the card on the surface of the bench in

4.4 Rates of glucose absorption from different regions of the alimentary canal in a freshly-killed mammal

TIME 2–2½ h

Absorption of glucose from the alimentary canal of a living mammal is believed to occur by an active process. In a freshly-killed mammal, however, it is generally possible to demonstrate only that glucose diffuses from different regions of the alimentary canal at different rates. The technique used is to introduce a solution of glucose into sacs prepared from lengths of the alimentary canal, and to monitor the rate at which glucose diffuses through the gut wall into a surrounding isotonic saline solution.

Preparation

The mammal should be killed immediately before it is required. If mammals such as rats or rabbits are used, a single animal will probably provide sufficient material for two students.

Materials

- Freshly-killed rat or rabbit
- Dissecting board and awls
- Dissecting instruments
- 30 cm³ 0.25 M glucose solution
- 50 cm³ 0.2 M sodium chloride solution
- Pie dish, or similar container, filled with 0.2 M sodium chloride solution
- 5 cm³ plastic syringe, fitted with a short needle
- 10 cm³ pipette, or plastic syringe
- Four test-tubes in a rack, or four flat-bottomed tubes, approximately 2 × 8 cm
- Clinistix reagent strips
- Cotton
- Glass-marking pen

Table 5 *Concentrations of glucose shown by Clinistix reagent strips.*

Colour of strip	Mass glucose (g) per 100 cm³ solution
Negative (pink)	0
Light (red-purple)	0.25
Medium (purple)	0.50
Dark (blue-purple)	0.75

Method

1 Dissect the mammal and remove the following regions from the alimentary canal.
 (i) The stomach, together with a 2 cm length of oesophagus at the anterior end and a 2 cm length of duodenum at the posterior end.
 (ii) A 10–15 cm length of duodenum, cut from below the stomach.
 (iii) A 10–15 cm length of ileum, cut anteriorly to the caecum.
 (iv) A 10–15 cm length of the large intestine, cut anteriorly to the anus.
 After excising each of these regions transfer them to a bath containing 0.2 M sodium chloride solution. Gently squeeze each region of the gut to expel the contents.

2 Tie the oesophageal end of the stomach with cotton, closing the lumen. Using a 5 cm³ syringe, fitted with a short needle and filled with the glucose solution, inject the solution into the cavity of the stomach, then tie off the duodenal end of the stomach with cotton. Wash off any surplus glucose that may have spilled over the surface of the stomach in the saline bath, then transfer the stomach to one of the tubes containing 10 cm³ 0.2 M sodium chloride solution. Label the tube.

3 Repeat procedure 2 with the duodenum, ileum and large intestine.

4 At intervals of 15 minutes, over a period of 90 minutes, test the saline surrounding each region of the alimentary canal for the presence of glucose, using the Clinistix reagent strips provided. Apply the test according to the manufacturer's instructions, and assume that the colour code corresponds to the concentrations of glucose shown in Table 5.
 Record your results in the form of a table. Plot your results as a graph.

5 How do you account for your results?

6 Describe any aspects of the experiment which were not controlled.

Topics to investigate

1 Transport of cations and anions across the small intestine of a freshly-killed mammal.

4.5 To determine the amount of urea in urine

TIME $1\frac{1}{2}$–2 h

A biochemical method is described for the estimation of urea in urine. The enzyme urease catalyses the reaction:

Urea $\longrightarrow$ Carbon dioxide $\longrightarrow$ Ammonium carbonate + Ammonia

Ammonium carbonate, the alkaline end-product of the reaction, can be estimated by titration with a standardised solution of hydrochloric acid, using methyl orange as indicator.

Preparation

It is advisable to analyse an early morning sample of urine in which amounts of urea are generally higher than in samples passed later in the day. The solution of urea required for this exercise contains 8 g solute dissolved in 100 cm³ water.

Materials

- Sample of urine
- 200 cm³ urea solution
- Urease tablets[1,3,4]
- 500 cm³ distilled water
- 500 cm³ 0.1 M hydrochloric acid
- Methyl orange indicator
- Burette
- Two 5 cm³ plastic syringes
- 25 cm³ pipette
- Six 250 cm³ conical flasks, fitted with bungs
- Pestle and mortar
- Water-bath, maintained at 35 °C

Method

1 Place two urease tablets, previously crushed, into a conical flask. Add 50 cm³ of distilled water, followed by 2 cm³ urea solution. Place a bung in the conical flask, so that it is air-tight, and transfer it to a water-bath maintained at 35 °C.
2 Make dilutions of the urea solution to give solutions containing, respectively, 4.0, 2.0, 1.0 and 0.5 g urea/100 cm³. Repeat procedure 1 with each of the solutions, and transfer the conical flasks to a water-bath maintained at 35 °C.
3 Repeat procedure 1 with the sample of urine, transferring the conical flask, with a rubber bung in position, to a water-bath at 35 °C.
4 After an incubation period of one hour, remove all the conical flasks and titrate the ammonium carbonate produced against 0.1 M hydrochloric acid, using methyl orange as indicator.
   Plot a calibration curve to show the volume of 0.1 M hydrochloric acid required to neutralise ammonium carbonate resulting from the breakdown of the urea. Use the calibration curve to estimate the percentage of urea in the urine sample. Give your result.
5 Comment briefly on your graph and on your results.

Topics to investigate

1 Urease activity in the spleen, liver and red blood corpuscles. (Urease activity is associated with the spleen, liver and red blood corpuscles.)
2 Urease activity in seeds of jack bean (soya bean) and water melon.
3 Urease activity of soil samples. (Microorganisms in the soil produce urease.)
4 Protein consumption and urea output.

4.6 Self assessment of cardiac and renal efficiency

TIME 3–3½ h

You are invited to investigate the way in which your cardiovascular system responds to a short period of vigorous exercise, and to examine the osmoregulatory and excretory functions of your kidneys.

Preparation

A bench suitable for use in this exercise is generally available from the school gymnasium.

Materials

- Bench, or stool, 25–35 cm in height
- Metronome
- Stop-watch, or clock with a second hand
- 250 cm³ measuring cylinder;
- Quantab chloride titrator no. 1176[4]

 WARNING: Those who are known to suffer from any serious defect of the cardiovascular or renal systems are advised against attempting this exercise.

Method

1 Record the pulse rate at the radial artery at two minutes, and again at one minute before commencing the following exercise. Step on and off the bench or stool 30 times a minute, in time with the metronome. In each movement lead with one foot, place both feet on the bench, return one foot to the floor, then both feet. Continue with the exercise for three minutes. Record the pulse rate immediately after exercising and at intervals of one minute for at least five minutes after cessation of the exercise. Record your results as a table. Plot your results as a graph and comment briefly on your results.

2 What would you conclude if your maximal pulse rate had been (i) higher and (ii) lower than the figure recorded on a previous occasion?

3 What are the chief limitations of the method employed in this test and how might they be overcome?

4 Empty the bladder. Record the volume of urine passed and measure the concentration of sodium chloride in the urine, using the Quantab chloride titrator.

5 Drink a measured volume of water between 750 cm³ and 1200 cm³, recording the volume of water consumed. After drinking the water, attempt to empty the bladder at intervals of 30 minutes, over a period of 3 hours. Measure the volume of urine in each sample and the concentration of sodium chloride. Record your results in the form of a table. Present your results as a graph.

6 What conclusions do you draw from your results?

Topics to investigate

1 Investigations of the cardiac cycle in a frog.
2 Electrocardiography.
3 Effects of exercise on (i) pulse rate and (ii) blood pressure.
4 Daily variations in the heart rate.
5 Daily variations in (i) the volume of urine produced and (ii) the sodium chloride content of the urine.
6 Does celery have a diuretic effect?

4.7 The effect of stimulants and depressants on the rate of heart beat in water fleas

TIME 2–2½h

If water fleas are viewed beneath a microscope, the hearts of these animals are visible through the translucent cuticle. The addition of a cardiac stimulant or depressant to the water surrounding the animals affects the heart, and changes in the rate of heart-beat can be observed and monitored.

Preparation

Solutions of adrenaline, chlorpromazine and salicylic acid each contain 0.1 g solute dissolved in 100 cm³ water. The ethanol–water mixture contains 1 cm³ ethanol added to 100 cm³ water.

Materials

- Culture of living water fleas (*Daphnia pulex*) in a pie dish or petri dish
- Five dropping pipettes, with rubber teats
- Four cavity slides and cover-slips
- 5 cm³ distilled water
- 5 cm³ ethanol/water mixture
- 5 cm³ adrenaline solution[1]
- 5 cm³ chlorpromazine solution[7]
- 5 cm³ salicylic acid solution[1,3,4]
- Stop-watch, or clock with a second hand
- Filter paper
- Glass-marking pen
- Compound microscope, fitted with low-powered lenses and objectives

Method

1 Use one of the pipettes to transfer 1–3 water fleas from the dish to a cavity slide. Remove the water from around the water fleas with filter paper, then add 1–2 drops of distilled water before covering the animals with a cover-slip. Transfer the slide to the stage of your microscope, and view the largest water flea, focusing on its heart, which lies in the anterior–dorsal region of the body. Tap your finger on the bench in rhythm with the heart-beat. Use the stop watch to record the number of beats per minute. Record the heart rate at intervals of two minutes, over a period of 10 minutes.
2 Repeat procedure 1 with other water fleas taken from the dish, clean slides and cover-slips and ethanol, adrenaline, chlorpromazine and salicylic acid. Record your results in the form of a table. Plot your results as a graph.
3 Describe your results and attempt to account for the effects of each treatment on the rate of heart-beat.

Topics to investigate

1 Experiments on the rate of heart-beat in (i) tadpoles of *Xenopus*, (ii) prawns or shrimps and (iii) lugworms.
2 Effects of temperature on the rate of heart-beat in invertebrate animals.
3 Effects of alcohols on the rate of heart-beat in water fleas.

The vertical distribution of water fleas is affected by environmental factors. This exercise involves an investigation of some of those factors, and the influence they exert on the behaviour of the animals.

Preparation

The apparatus, illustrated in Fig. 21, is assembled from a 10 cm^3 plastic syringe, rubber tubing and a 40 cm length of glass tubing, 4–6 mm in diameter. Prepare the suspension of yeast cells by adding 0.5 g dried yeast to 100 cm^3 water.

Materials

- Culture of living water fleas (*Daphnia pulex*) in a petri dish
- Apparatus
- Boss and clamp
- 5 cm^3 suspension of yeast cells
- 5 cm^3 olive oil
- Bench lamp, fitted with a 60 W bulb
- Black paper
- Scissors
- Adhesive tape
- Glass-marking pen

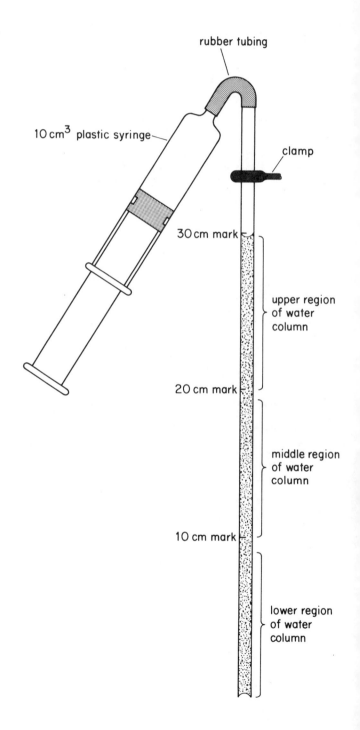

Fig. 21

Method

1 After assembling the apparatus, use the glass-marking pen to mark distances of 10, 20 and 30 cm from the open end of the glass tubing. Depress the plunger of the syringe. Introduce the open end of the glass tubing into the culture of living water fleas and raise the plunger to draw a 30 cm column of water, containing living animals, into the tubing. Clamp the apparatus, with the tubing in a vertical position, above the bench surface. The apparatus should appear as illustrated in Fig. 21.

Count the number of *Daphnia* in the upper, middle and lower regions of the water column. Make further counts at intervals of 5 minutes, over a period of 15 minutes. Record your results in the form of a table. Either describe, or draw, the distribution of animals after 15 minutes, and suggest possible reasons for this distribution.

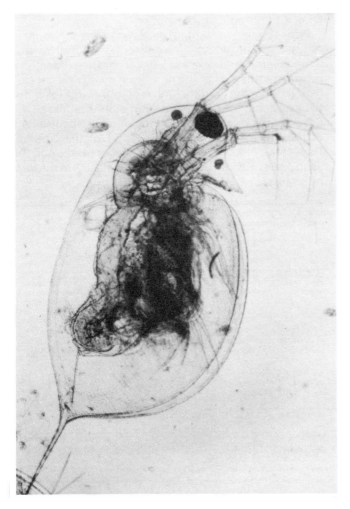

Daphnia × 100.

2 Make a sleeve out of the black paper, to cover the upper 10 cm region of the water column. Illuminate the lower region by means of the bench lamp, positioned at 10 cm from the glass tubing. What effect does this treatment have on the distribution of the animals after 5, 10 and 15 minutes?

3 Remove the apparatus from the clamp, insert the open end of the glass tubing into the suspension of yeast cells and draw in a column approximately 1 cm in length, in contact with the base of the water column.

Return the apparatus to the clamp. Observe, record and comment on any effect this treatment may have on the distribution of the water fleas.

4 Remove the apparatus from the clamp, eject the suspension of yeast cells and draw in a 1 cm column of olive oil, at the base of the water column and in contact with it. Return the apparatus to the clamp. Observe, record and comment on any effect this treatment may have on the distribution of the water fleas.

5 Give a general summary of your conclusions, pointing out any limitations of the investigation and indicating any areas in which further research may be necessary.

Topics to investigate

1 Factors affecting the distribution and orientation of planarians in water.
2 Responses of water skaters to directional illumination.
3 Direction-finding in pond snails.
4 Case-building and behaviour in caddis worms.

4.9 An investigation into behaviour of maggots of the blow-fly

TIME 2–2½ h

This exercise is concerned with factors influencing the rate and direction of locomotion in maggots of the blow-fly.

Preparation

The apparatus for this exercise is constructed from a 14 cm petri dish (or 9 cm petri dish) and a strip of stiff cardboard. Two or three days before the exercise is presented to students, paint one half of the petri dish–both the base and the lid–with at least two coats of black matt or gloss paint, applied to the outside of the dish. A cardboard partition, to completely separate the painted from the unpainted half of the dish, should be cut from stiff card. A central rectangular piece, measuring 0.5 cm × 1.5 cm should be cut from the central region of the partition, as illustrated in Fig. 22. It is important to ensure that this partition, when introduced into the petri dish and placed between the painted and unpainted halves, maintains its position and effectively separates the painted from the unpainted half, apart from the central hole.

Materials

- Blow-fly maggots (35–40) in a petri dish
- Apparatus shown in Fig. 20
- Two wooden blocks, each 15 cm in length, cut from 2 cm × 2 cm timber
- Stop-watch
- Bench lamp, fitted with a 75 W or 100 W bulb
- Metre rule
- Single sheets of writing paper, plastic sheeting and fine sand paper
- Forceps.

Method

1 Place a maggot on the bench surface. Note that as it moves a dark region of the gut, positioned just behind the head, pulsates backwards and forwards. Attempt to measure the rate of locomotion of a maggot when the animal is travelling over the surface of: (i) writing paper, (ii) plastic sheeting and (iii) sand paper, and to determine if the rate at which the gut pulsates is directly related to the rate of locomotion.

Place the sheet of writing paper on the bench surface, stand the two blocks of wood on it, parallel to one another and approximately 1.5–2.0 cm apart. Position a maggot at one end of the corridor between the blocks, facing down the corridor, and record the time taken for the maggot to travel from one end of the corridor to the other. At the same time count the number of pulsations of the gut as the maggot moves down the corridor. Repeat the procedure using the sheet of plastic and the sand paper. Record all your results, logically, in the form of a table.

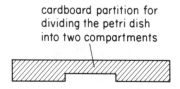

cardboard partition for dividing the petri dish into two compartments

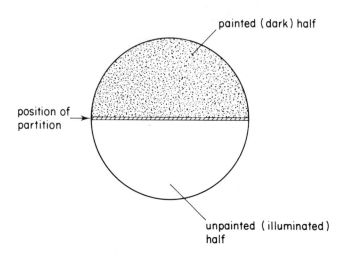

painted (dark) half

position of partition

unpainted (illuminated) half

Fig. 22 *Light–dark choice chamber*

2 Express rates of locomotion on the three different surfaces in centimetres per second.

3 Do you consider that the rate of locomotion is directly related to the number of pulsations of the gut? Give reasons for your answer.

4 Comment on the position of the maggot as it moved down the corridor. Name the response and state its adaptive significance.

5 Remove the right-hand block. Allow the maggot to move along the surface of the left-hand block. Either describe or draw the response of the maggot when it reaches the end of the wooden block. Replace the right-hand block, remove the left-hand block and repeat the procedure. Either describe or draw the response of the maggot.

6 Place the blocks approximately 2 mm apart, so that one side of each block is in lateral contact with the sides of the maggot's body. Allow the maggot to move down the central corridor and either describe or draw the response of the maggot when it reaches the end of the blocks.

7 Transfer 30 maggots to the apparatus and position them so that they occupy the unpainted half. Place the cardboard partition across the dish to separate the painted and unpainted halves, then replace the lid of the dish, so that the two halves are separated from one another. Switch off any bench lamps. At intervals of one minute, over a period of eight minutes, count the number of maggots in the painted (dark) and unpainted (illuminated) compartments of the dish. Record your results. Plot your results as a graph.

8 A researcher suspects that the rate of locomotion in maggots is affected by light intensity. Using the materials provided, design an experiment to test this hypothesis. Write out your proposals under the headings: (i) materials and (ii) method. The account should outline procedures, listed in logical sequence.

Topics to investigate

1 Behaviour of larval and adult flour beetles.
2 Differences in behaviour between closely related species of insects. (An investigation of this nature could, for example, be carried out using the larval and adult stages of small cabbage white butterflies and large cabbage white butterflies, or with different species of flour beetle.)

A more elaborate experiment is used in an attempt to determine if the rate of locomotion in a maggot is influenced primarily by the intensity of light falling on it, or by the wavelength (colour) of light transmitted through filters.

Preparation

Each Cinemoid light filter used in this exercise should be labelled with its colour and number, as specified by the manufacturers[8].

Materials

- Ten living blow-fly maggots in a container
- Cinemoid light filters Nos 1, 6, 18, 21 and 24, each approximately 10 × 6 cm
- Piece of thin cardboard, approximately 10 × 6 cm
- Pane of glass, approximately 10 cm²
- 9 cm plastic petri dish
- Bench lamp, fitted with a 75 W, or 100 W bulb
- Boss and clamp
- Stop-watch, or clock with a second hand
- Forceps
- Glass-marking pen

Method

1 Select a large, active maggot and transfer it to the petri dish. Set up the bench lamp at 15 cm above the bench surface, supported by a boss and clamp. Place the petri dish containing the maggots on the bench surface, immediately beneath the lamp. Allow 1–2 minutes for the maggot to adapt to its new environment. Observe that the maggot moves around the perimeter of the dish, maintaining contact with the sides. Place the rectangle of cardboard over one half of the dish, then cover it with the glass plate, so that it is held in place and heat is prevented from reaching the maggot. Measure, record, and comment on the rates of locomotion of the maggot in the illuminated and shaded halves of the dish.

Blow fly maggots moving away from a light source (at the top of the picture).

2 You are provided with five Cinemoid light filters, each labelled with a colour and number. Place each of these filters, in turn, to cover one half of the dish. Measure and record the time taken for the maggot to move around one half of the dish in white light and when each of the filters is in position, as illustrated in Fig. 23. Present your results in the form of a table.

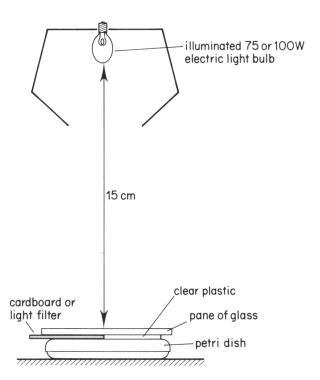

Fig. 23 *Apparatus for exercise 4.10*

3 Using the formula circumference $= \pi d$, calculate mean rates of locomotion, in centimetres per second, in white light, light of different qualities and in darkness.
4 Table 6 provides additional information about the light filters you have used. Plot graphs to show the relationships between rates of locomotion and (i) light intensity, (ii) light quality.
5 What conclusions do you draw and what further investigations could be carried out?

Table 6 *Characteristics of light filters.*

Light filter	Peak of transmittance (nm)	% absorbance
No. 1 Yellow	572	11.9
No. 6 Primary red	612	27.3
No. 18 Light blue	498	34.9
No. 21 Pea green	535	17.9
No. 24 Dark green	517	45.2
Cardboard	—	100.0

Topics to investigate

1 Responses of named invertebrate animals to light.
2 Light quality preferences of named photosynthetic motile green algae or bacteria.

4.11 Aspects of behaviour in the periwinkle

Periwinkles are littoral shelled molluscs. This is an ecological study in which you are asked to investigate aspects of behaviour in the animals and to comment on the adaptive significance of this behaviour.

Preparation

Each student should be presented with living periwinkles that are of different sizes. After collection from the sea shore, the animals can be kept alive for several days in covered buckets containing a little sea water and a supply of *Fucus serratus*.

Materials

- Thirty living periwinkles (*Littorina littorea*)
- White enamel dish, approximately 25 × 20 × 5 cm
- Cardboard box (to cover the dish)
- Four 100 cm³ measuring cylinders
- 500 cm³ beaker
- Spring balance, graduated to 1N in units of 0.05 N
- 1 m 15 A fuse wire
- 1 m 30 A fuse wire
- Bench lamp, fitted with a 60 W bulb
- 1.5 litres sea water
- 1.5 litres distilled water
- Black paper
- Adhesive tape
- Scissors
- White enamel paint
- Small paint brush
- Ruler, graduated in millimetres
- Top-pan balance

Method

1 Introduce five periwinkles into each of four numbered measuring cylinders. Add sea water up to the 100 cm³ mark in cylinders 3 and 4. Surround cylinders 2 and 4 with black paper, held in position with sticky tape. Observe and record movements of the animals in each cylinder over a period of 90 minutes. The animals in cylinders 2 and 4 may be observed by lifting the paper sleeve for a few seconds, then returning it to its original position.

Explain the aim of the experiment and discuss the adaptive significance of the behaviour shown by the periwinkles.

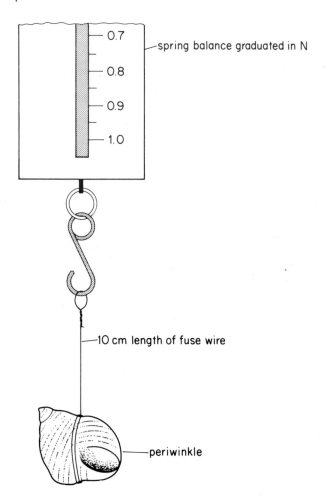

Fig. 24 *Method of attaching a periwinkle to a spring balance*

2 Place 10 periwinkles in the dish filled with sea water. Cover the dish with the cardboard box. Leave the periwinkles in darkness for approximately 5 minutes, then cut a hole, approximately 2 × 10 cm, in one side of the box, at 6–8 cm above the bench surface. Position the illuminated bench lamp at 5–10 cm from the hole. Cut a second hole in the roof of the box, through which movements of the animals can be observed. What do you observe after 10, 20 and 30 minutes? How might this behaviour help to promote the survival of the animal in its natural habitat?

3 Transfer 10 periwinkles to the beaker. Half fill the beaker with sea water and count the number of periwinkles that have emerged from their shells after being immersed for 10 minutes. Make dilutions of the sea water, repeat the test, and record your results in the form of a table. Plot your results as a graph. What do you conclude?

4 Transfer 10 periwinkles of different sizes to the dish, filled with sea water. Weigh each animal on the top-pan balance, number the molluscs on their shells with white paint, and record the mass of each animal.

   Cut the 15 A fuse wire (or 30 A fuse wire, depending on the mass of the animals) into 10 cm lengths. Wrap one end of a piece of wire around each periwinkle, as illustrated in Fig. 24, taking care to avoid obstructing the shell aperture. Make a small loop at the other end of the wire. As soon as a periwinkle emerges from its shell and begins to move, slip the hook of the spring balance through the loop. Apply sufficient force to pull the periwinkle from the substratum, and read off the force required to effect detachment from the substratum. Repeat this procedure with each periwinkle in turn, taking at least two readings for each animal. Record your results in the form of a table. Plot your results as a graph. Would you expect to find a positive correlation between the mass of an animal and the force required to dislodge it from the substratum? Give reasons for your answer.

5 After examining periwinkles collected from different regions of the shore, a researcher considered that those collected from just below the high-water zone had relatively thicker shells than those collected from around the low-water mark. How might this hypothesis be tested?

Topics to investigate

1 The adhesive power of limpets in relation to (i) body mass and (ii) position on the shore line.
2 Variations in the behaviour of different species of littoral periwinkles (e.g. *Littorina littorea*, *Littorina obtusata* and *Littorina neritoides*).
3 Aspects of behaviour in the dog whelk.
4 Shell mass to body mass ratios in shelled molluscs from different zones of the sea shore.

Periwinkles.

4.12 Locomotion and behaviour in larvae of the mosquito

TIME 1½–2 h, *June–August*

The exercise requires observation of aspects of locomotion and behaviour in aquatic larvae of the mosquito.

Preparation

Mosquito larvae may be obtained during the summer by leaving bowls or buckets of water in positions likely to be visited by mosquitoes. Generally, a culture obtained in this way will consist of larvae of different ages and sizes, as required for this investigation.

Materials

- Culture of living mosquito larvae
- Two 100 cm³ measuring cylinders
- Bench lamp, fitted with a 60 W bulf
- Cardboard, approximately 10 × 10 cm
- Ruler, graduated in millimetres

Method

1 Tip the culture of mosquito larvae into one of the measuring cylinders and, if necessary, add water to bring the water level to the 100 cm³ mark. Set up the bench lamp at one side of the measuring cylinder, to illuminate the larvae. Observe and describe, with the aid of simple drawings or diagrams, methods used by the larvae to (i) ascend and (ii) descend through the water. At one point in your observation, tap the side of the vessel and observe if this has any effect on the direction and method of locomotion.

2 Attempt to relate the size of individual larvae to (i) the number of swimming movements made per minute and (ii) the rate of locomotion.

3 Transfer two of the largest larvae to the second 100 cm³ measuring cylinder, filled with water to the 100 cm³ mark. Observe, and record on graph paper the relative amounts of time spent by each of the larvae on the surface and at different depths in the water. What do you conclude? What effect would you predict if the temperature of the water were to be raised by 10° C?

4 Add a further 4–5 large larvae to those in the second measuring cylinder. Allow approximately one minute for the larvae to aggregate at the surface, then place the cardboard square on top of the measuring cylinder, vibrating the glass as the square is placed in position. Remove the cardboard square after 10 seconds. Repeat this procedure 10 times at intervals of two minutes. Observe, and record on graph paper, movements of the larvae in response to this stimulus. What do you conclude? What response would you expect if the stimulus was repeated again after a resting period of 20 minutes?

Topics to investigate

1 Locomotion in named insect larvae.
2 Habituation in invertebrates.

Aspects of osmoregulation and behaviour in the shore crab

The shore crab is a hardy littoral species, tolerant of a wide range of concentrations of sodium chloride in water. In this exercise some physiological and behavioural adaptations of the shore crab are investigated.

Preparation

Crabs used in this exercise should measure 4–6 cm across the carapace. Shore crabs can be kept in tanks of natural or artificial sea water, and if fed sparingly on a balanced diet, such as dog biscuits, will survive for a number of years. The beaker should contain a 5 cm depth of sand covered by a similar depth of sea water.

Materials

- Living shore crab (*Carcinas moenas*)
- 100 cm³ distilled water
- White enamel dish, approximately 25 × 20 × 5 cm
- Strip of black paper, 25 × 5 cm
- Six flat-bottomed tubes, each approximately 2 × 8 cm
- Litre beaker, containing sand and water
- 100 cm³ measuring cylinder
- 5 cm³ plastic syringe
- Jam jar (or similar container, which will retain the crab)
- Quantab chloride titrator no. 1175.[4]

A Shore crab.

Method

1 Place the crab in the jam jar and add 100 cm³ distilled water. At intervals of 10 minutes, over a period of 60 minutes, withdraw 2 cm³ water from the jar and measure the mass of sodium chloride in the water, together with the concentration of chloride ions. Use a Quantab chloride titrator no. 1175 and the flat-bottomed tubes when making these measurements. Record your results in the form of a table.

2 From your results, plot a graph that will enable you to predict the time taken by the crab to establish an ionic equilibrium with the surrounding water. Explain the construction of your graph.

3 What are the chief sources of error in the method used to estimate concentrations of sodium chloride?

4 Transfer the crab to the litre beaker containing sand covered by water. Describe in detail the behaviour you observe and attempt to account for the adaptive significance of this behaviour.

5 Transfer the crab to the enamel dish and place the strip of black paper along one side of the dish. Describe the behaviour you observe and attempt to account for the adaptive significance of this behaviour.

6 Remove the black paper from the dish. Position the crab centrally. Tap the posterior part of the carapace with a pen or pencil, observing the manner and direction in which the crab moves. Tap the base of the carapace at least ten times, recording the direction of movement.

   Fill the dish with sea water or tap water and repeat this procedure. Record all of your results and comment on them.

Topics to investigate

1 Osmoregulation in the edible crab and fiddler crab.
2 Behaviour in the edible crab and fiddler crab.
3 Osmoregulation and behaviour in the lugworm.

4.14 Removal of organ systems from a rat

Routine dissections of a mammal lie outside the scope of this book. Even so, the final exercise is included as it demands considerable care, patience and manual dexterity in those who already have some proficiency in dissection.

Preparation

The exercise may be carried out on a freshly-killed or preserved animal, although removal of intact regions from the alimentary canal of a preserved animal may present considerable difficulties. The saline solution contains 9 g sodium chloride dissolved in each litre of water.

Materials

- Freshly-killed male rat
- Dissecting board and awls
- Dissecting instruments
- Large dish containing black wax
- Litre of saline solution, supplied in a beaker
- Stiff cardboard, approximately 7 × 10 cm
- 20–30 large dissecting pins

Method

1 Remove, by careful dissection from the rat provided, each of the following organ systems.
   (i) The respiratory system.
   (ii) The stomach, duodenum, pancreas and liver.
   (iii) The large intestine.
   (iv) Either the excretory system or the reproductive system.
2 After dissecting out each system, lift it gently on to the cardboard tray provided and transfer it to the dish of black wax. Pin out the dissection and cover it with saline solution.
3 Make large, labelled drawings of each organ system.

Topics to investigate

1 Removal of an intact alimentary canal from a freshly-killed cockroach or locust.

Results

1.1 A colorimetric method for the estimation of glucose (or reducing sugars) in solution

1, 2 and 3 See Table 7 and Fig. 25.

Table 7

Concentration of glucose (g/100 cm³)	Time for reduction of manganate(VII) (seconds)
10	105
9	115
8	135
7	147
6	172
5	210
4	260
3	345
2	510
1	870

4 In trials the following concentrations of glucose were used:
A = 7 g/100 cm³, B = 10 g/100 cm³, C = 2 g/100 cm³, D = 4 g/100 cm³, E = distilled water.

5 The graph is curvilinear and the rate at which the manganate (VII) is reduced is directly related to the amount of glucose in solution, increasing as the concentration of glucose is increased. At low concentrations of glucose, however, the rate of reduction is slow and the end point, at which the mixture changes from translucent pink to colourless, often difficult to detect. Unfortunately, the reaction is not specific to glucose, but may occur as the result of other reducing agents, such as fructose, maltose or vitamin C, being present.

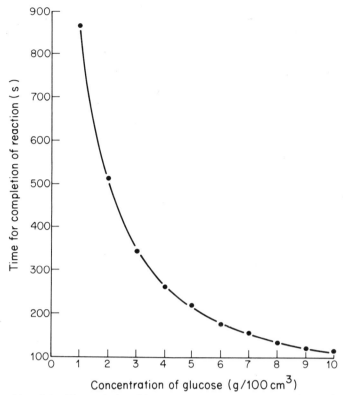

Fig. 25 *The relationship between the concentration of glucose and the time taken for reduction of a standard solution of potassium manganate(VII)*

1.2 Chemical methods for estimating the pH, hardness, salinity and oxygen content of water

1 The pH of pond water used in trials = 6.3.
The pH of sea water used in trials = 7.3.
The results obtained are given in Table 8 and plotted in Fig. 26.

Table 8

Volume 0.1 M HCl added (cm³)	pH	
	Pond water	Sea water
0	6.3	7.3
1	5.9	7.2
2	5.5	7.0
3	5.2	6.9
4	4.9	6.5
5	4.6	6.1
6	4.1	5.5
7	3.8	5.3
8	3.6	5.0
9	3.3	4.7
10	3.0	4.3

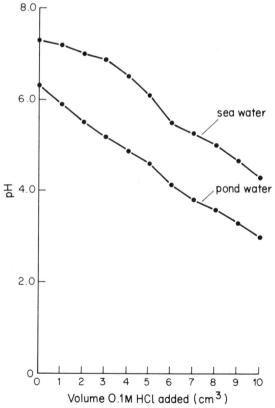

Fig. 26 *Changes in the pH of pond water and sea water as a result of adding 0.1M HCl*

The pond water was found to be slightly acid, whereas sea water was alkaline. When small volumes of acid are added to water samples, ions in the sea water act as a buffer, causing a relatively slower fall in the pH than if similar amounts of acid are added to pond water.

2 Hardness of pond water used in trials = 110 p.p.m.
Hardness of sea water used in trials = 6700 p.p.m.

The hardness of water results from the presence of dissolved salts of calcium (and magnesium), chiefly in the form of bicarbonate. Calcium is an essential nutrient for plants, forming a component of the cell wall. Similarly, calcium is an essential nutrient for all animals, but it is of particular importance to shelled animals, such as molluscs and crustaceans, which may not be able to extract sufficient calcium from soft waters in order to construct adequate exoskeletons.

3 The sodium chloride content of pond water used in trials = 0.054 g/100 cm^3.
The sodium chloride content of sea water = 3.4 g/100 cm^3.

Both sodium ions (Na^+) and chlorine ions (Cl^-) are essential for the maintenance of health in plants and animals. Pond plants and pond animals, surrounded by low concentrations of Na^+ and Cl^- ions, must be able to absorb these ions against a concentration gradient. Additionally, they are in constant danger of losing these ions to the surrounding water. Conversely, plants and animals living in the sea must possess some mechanism for excluding, or excreting, these ions from their bodies. As concentrated solutions of sodium chloride exert an osmotic effect, marine organisms must be able either to resist, or to counteract, removal of water from their bodies.

4 Volume 0.001 M solution of sodium thiosulphate added to boiled pond water = 3.6 cm^3.
Hence, approximate percentage of oxygen = 0.0432%.
(This figure may, however, in part be due to oxidising agents in the water.)
Volume 0.001 M solution of sodium thiosulphate added to unboiled pond water = 18.2 cm^3.
Hence, approximate percentage of oxygen = 0.2184%.
These results show that the pond water, maintained in contact with the air for a period of 24 hours, was capable of absorbing further oxygen from the air to increase its percentage of saturation by 0.1752%.

5 Volume 0.001 M solution of sodium thiosulphate added to boiled sea water = 7.8 cm^3. Hence, approximate percentage of oxygen = 0.0936%.
This result is almost certainly elevated by the presence of oxidising agents such as Cl^- ions in the water.
Volume 0.001 M solution of sodium thiosulphate added to unboiled sea water = 17.3 cm^3.
Hence, approximate percentage of oxygen = 0.2076%.
Results show that a similar volume of sea water, maintained under identical conditions to that of pond water, absorbs further oxygen from the air to increase its percentage of saturation by 0.114%.
Generally, concentrations of oxygen are found to be lower in sea water than in pond water, an effect that is related to the presence of larger concentrations of solutes in sea water.

Chemical methods of estimating the oxygen content of water samples are often inaccurate, as much depends on the conditions of storage and treatment of the water after it has been collected. Additionally, when titrations are made against sodium thiosulphate, results are affected by the presence of oxidising agents in the water.

1.3 The interconversion of nitrogenous compounds in a loam soil.

1 and 2 See Table 9.

Table 9

Beaker no.	Concentration of ions (p.p.m.) at 6-hourly intervals														
	NO_2^-					NO_3^-					NH_4^+				
	0	6	12	18	24	0	6	12	18	24	0	6	12	18	24
1	0	4	4	4	4	90	90	90	90	90	30	30	30	30	30
2	0	4	20	20	40	90	90	120	120	120	1600+				→
3	0	4	4	4	4	90	90	90	90	90	1600+				→
4	200+				→	90	120	500	2000+	→	30	30	30	30	30
5	0	20	40	100	200	2000+				→	30	30	30	30	30
6	0	4	4	4	4	90	90	90	90	90	30	30	40	40	120
7	0	40	100	100	200	90	90	90	90	90	30	40	240	400	800

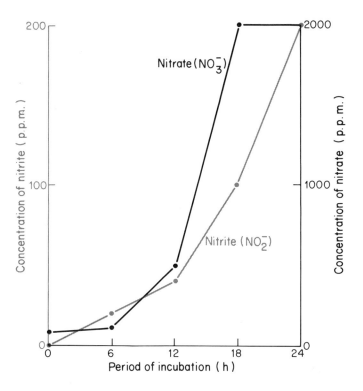

Fig. 27 *Rate of formation of nitrate from nitrite in aerated soil, and of nitrite from nitrate in waterlogged soil*

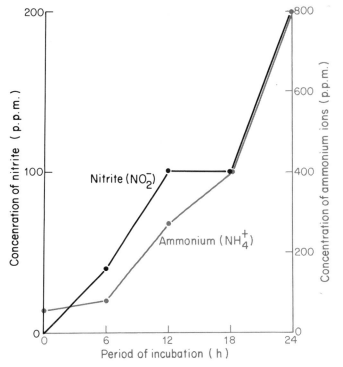

Fig. 28 *Production of nitrite and ammonia following incubation of a loam soil with a solution of urea*

3 See Fig. 27.

4 See Fig. 28.

5 Green plants are capable of absorbing NO_3^- and NH_4^+ ions from the soil, and they are dependent on soil micro-organisms for the formation of these compounds. Other micro-organisms, however, form NO_2^- ions in the soil; these nitrite ions are toxic to green plants.

Beaker 1 In a waterlogged, anaerobic soil there is a slow conversion to NO_3^- to NO_2^-, chiefly as a result of the activity of denitrifying bacteria. Even so, levels of nitrite at something less than 10 p.p.m. probably have little adverse effect on the growth of green plants.

Beaker 2 When ammonia is present in an aerated soil, there is a slow oxidation of the ammonia first to nitrite, then to nitrate. This results from the activity of nitrifying bacteria such as *Nitrosomonas*, which forms nitrite from ammonia, and then *Nitrobacter*, which forms nitrate from nitrite.

Beaker 3 When ammonia is added to a waterlogged soil there is little, if any, oxidation of this compound to nitrite and nitrate.

Beaker 4 In a moist, aerated soil there is rapid conversion of NO_2^- to NO_3^-, resulting from the activity of aerobic nitrifying bacteria, such as *Nitrobacter*.

Beaker 5 In a waterlogged soil, under anaerobic conditions, NO_3^- in the soil is slowly converted to NO_2^-. If appreciable amounts of nitrate, for example in the form of artificial fertilisers, are added to a waterlogged soil, then appreciable amounts of nitrite may be formed. This may have a toxic effect on plants and lead to spoilage of a crop.

Beaker 6 Protein, mixed with soil, undergoes slow decomposition, a process sometimes known as 'mineralisation' or 'ammonification'. Ammonia is released to the atmosphere.

Beaker 7 Urease, produced by soil micro-organisms, decomposes urea into ammonia and carbon dioxide.

$$Urea \xrightarrow{\text{Urease}} Ammonia + Carbon\ dioxide$$

In addition to the formation of NH_4^+ ions, large amounts of nitrite are also formed, especially under anaerobic conditions.

1.4 A systematic identification of carbohydrates

1 See Fig. 29.
2 The carbohydrates used in trials were: A = starch, B = maltose, C = fructose.
3 Providing it was clearly established that one powder was starch and the other cellulose, each powder could be added to an aqueous solution of amylase plus maltase. After incubating the mixture at a suitable temperature, Benedict's test for reducing sugars and the Clinistix test for glucose could be applied. Positive results would identify starch, hydrolysed by these enzymes to reducing sugars, including glucose.

$$Starch \xrightarrow{Amylase} Maltose \xrightarrow{Maltase} Glucose$$

Cellulose, on the other hand, would not be hydrolysed by these enzymes. Therefore, in the case of cellulose, tests for reducing sugars would prove negative.

1.5 Estimation of citric acid, ascorbic acid and reducing sugars in the juice of orange, lemon and grapefruit.

1 See Table 10.

Table 10

Fruit	Total mass (g)	Volume juice (cm³)	% Juice
Orange	158.5	46.0	29.02
Lemon	97.5	37.0	37.9
Grapefruit	357.0	121.0	33.9

2 Volume of citric acid solution required to neutralise 20 cm³ sodium hydroxide solution = 6.5 cm³.
Volume of orange juice required to neutralise 20 cm³ sodium hydroxide solution = 14.6 cm³.
Volume of lemon juice required to neutralise 20 cm³ sodium hydroxide solution = 1.8 cm³.
Volume of grapefruit juice required to neutralise 20 cm³ sodium hydroxide solution = 8.4 cm³.
Total mass of citric acid in 100 cm³ of each juice: orange = 0.45 g, lemon = 3.61 g, grapefruit = 0.78 g.

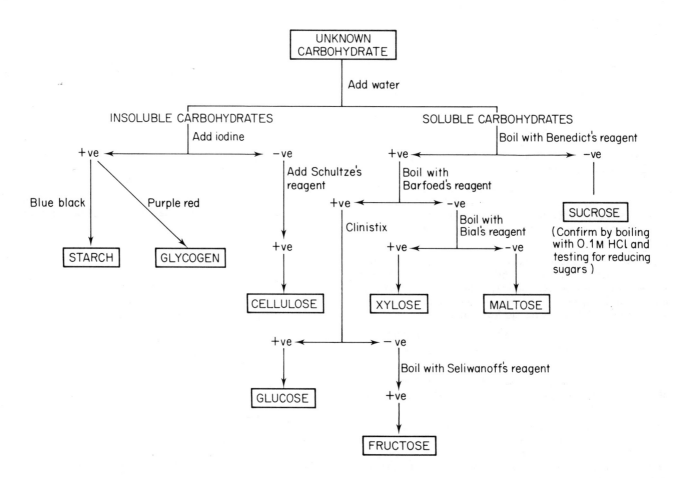

Fig. 29 *A scheme for the identification of carbohydrates*

3 See Table 11.

Table 11

| Fruit | Vol. juice containing 1 mg ascorbic acid (cm³) | |
	1st reading	2nd reading
Orange	2.5	2.5
Lemon	1.5	1.5
Grapefruit	2.0	1.9

  (i) Total mass of ascorbic acid in each fruit: orange = 17.7 mg, lemon = 24.67 mg, grapefruit = 63.7 mg.

 (ii) Total mass of ascorbic acid in 100 cm³ of each juice: orange = 38.5 mg, lemon = 66.6 mg, grapefruit = 52.6 mg.

4 See Table 12.

Table 12

| Dilution of juice | Approximate concentration of reducing sugars (g/100 cm³) | | |
	Orange	Lemon	Grapefruit
Undiluted (1.0)	2.0	2.0	2.0
0.5	2.0	1.0	2.0
0.25	1.5	0.5	1.0
0.125	0.75	0.25	0.5

Total mass of reducing sugars in 100 cm³ of each juice: orange = 6.0 g, lemon = 2.0 g, grapefruit = 4.0 g.

5 (i) It is difficult, when cutting fruits by hand, to ensure that all the juice and none of the solid material is collected. There is therefore a need, in this part of the experiment, to filter juice from cell debris.

 (ii) Other acids, in addition to citric acid, are present in the orange and other citrus fruits. The method employed to estimate citric acid therefore measures total acid, including ascorbic acid (vitamin C).

(iii) The chief limitation imposed by the use of DCPIP tablets to estimate ascorbic acid is that they are difficult to dissolve, and this may lead to a tendency to overestimate amounts of ascorbic acid present in samples.

(iv) Estimation of reducing sugars by a semi-quantitative method, which relies on visual comparisons, is unreliable. More accurate results could be obtained by the use of a colorimeter or, alternatively, by evaporating 100 cm³ juice to dryness, and weighing the residues, which are composed largely of reducing sugars.

1.6 Distribution of chemical compounds in plant tissues

1 A: (i) mesocarp, (ii) seed, (iii) placenta, (iv) epicarp, (v) vascular bundles, (vi) funicle.
B: (i) Carpel, (ii) placenta, (iii) seed, (iv) mesocarp, (v) epicarp.
C: (i) pericarp and testa, (ii) aleurone layer, (iii) endosperm, (iv) scutellum, (v) cotyledon, (vi) radicle, (vii) hilum, (viii) coleorhiza, (ix) secondary root, (x) plumule, (xi) coleoptile.
D: (i) cork, (ii) shoot, (iii) rhizome scar, (iv) vascular bundle, (v) adventitious root.

4 *Stain W* Iodine in a solution of KI, which produces a blue colour with starch and a purple colour with the hydrolysis products of starch, such as dextrins.
Stain X Silver nitrate, which is reduced to metallic silver in tissues containing glucose or other reducing sugars.
Stain Y Bromocresol green in propan-2-ol, which in an acid medium stains proteins green. If ponceau S was used, areas containing proteins are stained red.
Stain Z Tetrazolium salt, a vital stain, reduced by dehydrogenases to an intensely red formazan. The stain indicates regions of active mitochondria, such as those in meristematic tissues.

5 See Table 13.

Table 13

| Stain | Tissues stained | | | |
	Cucumber	Banana	Maize grain	Potato
W (iodine)	None	Outer region of mesocarp	Endosperm and root cap Dextrins in cotyledon	Tuber and shoots
X (silver nitrate)	Mesocarp, especially around seeds	Epicarp and centre of mesocarp	Cotyledon, shoot apex and root apex	Shoots Traces in tuber
Y (bromocresol green)	Outer region of mesocarp	Outer region of mesocarp	Aleurone layer of grain	Shoot apex Traces in tuber, especially beneath cork
Z (tetrazolium salt)	None	Outer region of mesocarp (in unripe fruits)	Shoot apex, root apex and embryo	Shoot apex

1.7 Identification and separation of photosynthetic pigments by thin-layer chromatography

7 See Fig. 30.

Solvent A

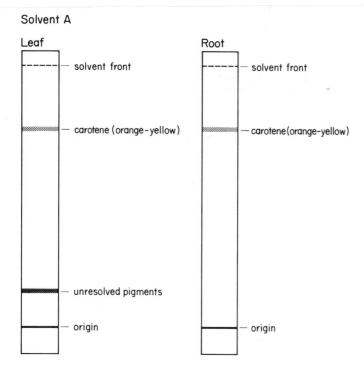

Solvent B

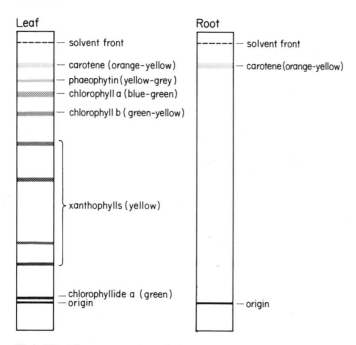

Fig. 30 *The appearance of chromatograms developed from leaf and root extracts of carrot, using two different mixtures of solvents*

8 (i) Solvent A, which effectively separates photosynthetic pigments when used with paper as an absorbent, fails to do so when used with silica gel as an absorbent. Even so, a carotene is separated from a mixture of xanthophylls and chlorophylls.

Solvent B effects efficient separation of photosynthetic pigments when used with silica gel as an absorbent.

(ii) Solvent B effects efficient separation of pigments from the leaf into at least eight different components, consisting of a single carotene, two chlorophylls and probably as many as four different xanthophylls. The root, however, contains an identical carotene, but no chlorophylls or xanthophylls.

9 See Table 14.

Table 14

Pigment	Colour of pigment	R_f value
Carotene	Orange-yellow	0.96
Phaeophytin	Yellow-grey	0.81
Chlorophyll a	Blue-green	0.75
Chlorophyll b	Green-yellow	0.70
Xanthophyll a	Yellow	0.51
Xanthophyll b	Yellow	0.37
Xanthophyll c	Yellow	0.17
Xanthophyll d	Yellow	0.08
Chlorophyllide a	Green	0.01

10 R_f values may be affected by: (i) temperature of the air, (ii) chemical nature of the solvent, (iii) chemical nature of the absorbent, (iv) pore size of the absorbent.

1.8 Identificatin and separation of anthocyanidins by paper chromatography

5 R_f values will depend upon the type of chromatographic paper used. When Whatman no. 1 paper was used in trials, the following R_f values were obtained:
Pelargonidin (red) = 69
Malvidin (mauve) = 61
Cyanidin (red) = 49
Petunidin (purple) = 46
Delphinidin (blue) = 32
6 R_f values of unidentified pigments were:
Specimen X = 69 = pelargonidin
Specimen Y = 49 = cyanidin
Specimen Z = 49 = cyanidin
Therefore the strawberry contains pelargonidin and both the apple and red cabbage contain cyanidin.

1.9 Chemical analysis of wines

1 The total acidity of a French white wine (wine A) = 91
 The total acidity of a pale 'cream' sherry (wine B) = 69
3 See Fig. 31.

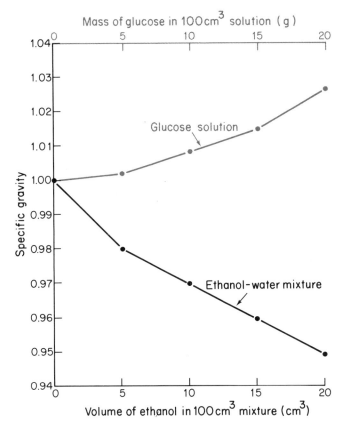

Fig. 31 *Specific gravity of glucose solutions and ethanol–water mixtures*

4 Wine A contained less than 5 g glucose/100 cm³ and approximately 10% ethanol by volume.
 Wine B contained from 5–10 g glucose/100 cm³ and from 15–20% ethanol by volume.
5 R_f values:
 Lactic acid = 0.88
 Succinic acid = 0.82
 Malic acid = 0.74
 Citric acid = 0.67
 Tartaric acid = 0.58
 Wine A contained lactic and citric acids. Wine B contained tartaric acid. Other acids may have been present but it was difficult to effect separation and therefore difficult to make an identification.

1.10 Estimation of the chlorophyll and starch content of leaves

5 See Table 15.

Table 15

Test material	% Absorbance
Water	50
Leaf 1	53
Leaf 2	56
Leaf 3	68
Leaf 4	63
Leaf 5	61
Leaf 6	54
Leaf 7	53
Leaf 8	51

6 and 7 See Table 16.

Table 16

Test material	% Absorbance
Iodine solution	65
Leaf 1	67
Leaf 2	67
Leaf 3	78
Leaf 4	71
Leaf 5	68
Leaf 6	68
Leaf 7	68
Leaf 8	67

8 See Fig. 32 on page 100.

9 Different levels of chlorophyll and starch were recorded in each leaf tested. Generally, the level of starch in a leaf was closely related to the amount of chlorophyll present, being highest in those leaves with most chlorophyll, and lowest in those leaves with least chlorophyll.
 Young leaves, near the apex of the stem, and old leaves, at the base of the stem, contained less chlorophyll and starch than leaves at nodes numbered 3, 4 and 5, positioned at approximately one-third of the total stem length, measured from below the apex.

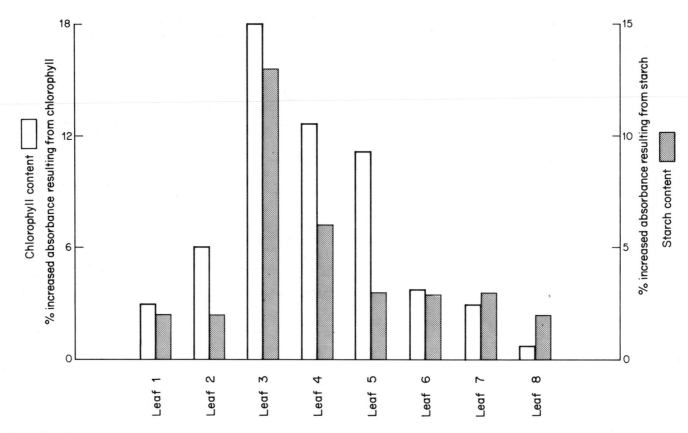

Fig. 32 *Relative amounts of chlorophyll and starch in the leaves of a cucumber plant*

2.1 To determine the rate of denaturation of a protein by heat, ethanol and copper(II) sulphate

2 See Table 17 and Fig. 33.

Table 17

Temperature of incubation (°C)	% absorbance of albumen			
	15 min	30 min	45 min	60 min
30	0	0	0	0
40	0	0	0	0
60	2	2	3	6
80	100	100	100	100

Incubation of egg albumen at 30 °C and 40 °C caused no increase in the percentage absorption, nor any visible signs of denaturation, over a period of 60 minutes. At 60 °C, however, slow denaturation was observed, while at 80 °C denaturation was rapid, being complete within the first 15 minutes of incubation. Further investigations need to be carried out to determine the rate of denaturation of the albumen at 65, 70 and 75 °C.

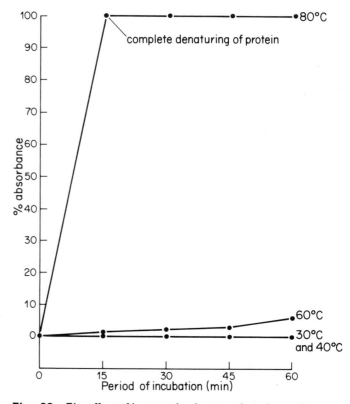

Fig. 33 *The effect of heat on the denaturation of egg albumen*

3 See Table 18 and Fig. 34.

Table 18

% ethanol in mixture	% absorbance of albumen
0	0
10	4
20	20
30	80
40	100
50	100

Ethanol is a toxic compound, which causes some denaturation of albumen at all concentrations. While the extent of the denaturation is relatively small at 10 % v/v ethanol, it increases through 20–30 % v/v ethanol, to become complete between 30–40 % v/v ethanol.

4 Introduce 5 cm³ egg albumen and 5 cm³ 0.1 M copper(II) sulphate solution (= 0.05 M copper(II) sulphate solution) into a flat-bottomed tube, gently rock the mixture to ensure that mixing has taken place, and measure the absorbance of the mixture. Repeat this procedure using 0.025 M, 0.0125 M, 0.00625 M, 0.003125 M and 0.0015625 M copper(II) sulphate solutions. Record your results in the form of a table (Table 19) and present your results as a graph (Fig. 35).

Table 19

Molarity of CuSO$_4$ in mixture	% absorbance of albumen
0.05	100
0.025	100
0.0125	100
0.00625	90
0.003125	60
0.0015625	10

Copper(II) sulphate causes complete denaturation of the egg albumen at a concentration between 0.00625 M and 0.0125 M. Further dilutions of the 0.1 M copper(II) sulphate solution would have to be made in order to determine the lowest concentration of copper(II) sulphate that would cause complete denaturation of the egg albumen.

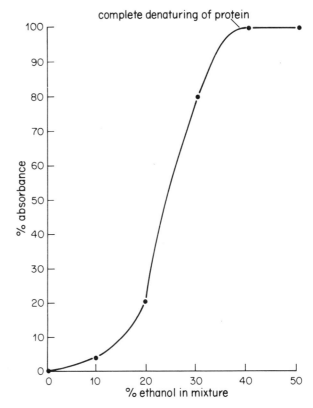

Fig. 34 *The effect of ethanol on the denaturation of egg albumen*

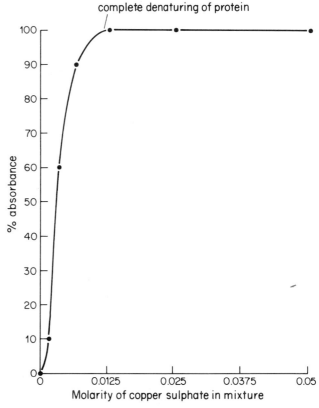

Fig. 35 *The effect of copper(II) sulphate on the denaturation of egg albumen*

101

2.2 Losses of protein and sugars from foods as a result of cooking, freezing and drying

1 See Tables 20 and 21 for results for fresh green beans.

Table 20

Period of boiling (min)	Protein content of cell sap (mg/100 cm³)	Protein content of water (mg/100 cm³)
0	100	0
15	100	trace
30	100	trace
45	30	trace
60	30	30

Table 21

Period of boiling (min)	Glucose content of cell sap (mg/100 cm³)	Glucose content of water (mg/100 cm³)
0	45	0
15	25	0
30	trace	trace
45	trace	trace
60	trace	trace

Protein content of frozen sliced beans = 100 mg/100 cm³
Glucose content of frozen sliced beans = trace

2 See Tables 22 and 23 for results for fresh carrots.

Table 22

Period of boiling (min)	Protein content of cell sap (mg/100 cm³)	Protein content of water (mg/100 cm³)
0	30	0
15	trace	trace
30	trace	trace
45	trace	trace
60	trace	trace

Table 23

Period of boiling (min)	Glucose content of cell sap (mg/100 cm³)	Glucose content of water (mg/100 cm³)
0	90	0
15	45	trace
30	trace	trace
45	trace	trace
60	trace	trace

Protein content of frozen carrots = trace
Glucose content of frozen carrots = trace

3 Protein content: fresh sultana grapes = trace
dried sultana grapes = nil

Glucose content: fresh sultana grapes = 130 mg/100 cm³
dried sultana grapes = 90 mg/100 cm³

4 See Tables 24 and 25 for results for grass.

Table 24

Time after harvesting (days)	Protein content of grass (mg/100 cm³)
0	100
1	30
2	30
3	30
4	30
5	trace
6	trace
7	trace

Table 25

Time after harvesting (days)	Glucose content of grass (mg/100 cm³)
0	45
1	trace
2	trace
3	trace
4	trace
5	trace
6	trace
7	trace

5 The results demonstrate that as a result of cooking, both protein and sugars are leached out of food into the surrounding water. As the period of cooking is increased, so increasing amounts of protein and sugars are removed from the food into the surrounding water. After cooking vegetables for one hour, for example, most of the protein and sugars have passed into the water, which is generally discarded when the vegetables are removed from the pan.

Frozen vegetables contain less protein and sugars than fresh vegetables. These losses, however, may result primarily from cooking before the material is deep-frozen.

During the drying of fruits there is an appreciable loss of protein, but relatively little loss of glucose, providing the fruit is kept in an air-tight container.

Similarly, when grass is converted into hay there is an appreciable loss of protein during drying of the crop, and quite a marked fall in the level of sugars. Hence the nutritive value of dried grass may be appreciably less than that of fresh grass.

As a result of these experiments the following recommendations might be made:

(i) Wherever possible, vegetables should be eaten in the raw state. Alternatively they should be cooked for a relatively short period of time only, to avoid unnecessary losses of nutrients.

(ii) More nutrients appear to be preserved by deep-freezing than by drying.

(iii) Farmers should be made aware that hay is less nutritious than fresh grass. Where hay is fed to cattle, there is a need to supply an additional protein-rich food. Alternatively, farmers should be encouraged to make silage rather than hay, as more nutrients, especially protein, are preserved by the process of silage-making.

2.3 A method for measuring the density of milk: copper proteinate formation

1 and 2 See Table 26.

Table 26

Composition of mixture		Time for descent of drop (s)			
% milk	% water	1st run	2nd run	3rd run	Mean value
100	0	14	13	14	13.6
90	10	18	19	19	18.6
80	20	29	30	28	29.0
70	30	75	78	73	75.3
60	40	124	128	126	126.0
50	50	142	144	149	145.0
40	60	floats	floats	floats	floats

3 The sample of milk used in trials had the same density as the solution of copper(II) sulphate when mixed with water in the proportions 48% milk: 52% water. After observing that a milk–water mixture containing 50% v/v milk sank, whereas one containing 40% v/v milk floated, dilutions were made to give volumes of milk between 40–50% v/v in each mixture. At a density equivalent to that of 0.1 M copper(II) sulphate, the drop remained in position, neither sinking nor floating.
4 See Fig. 36.
5 The chief source of error in this experiment is variation in the size of the drop that is released from the syringe. The problem could be overcome by using a syringe fitted with a device that delivers drops of uniform size.

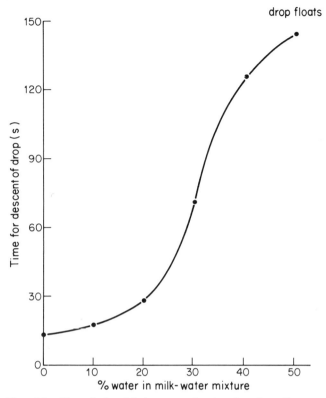

Fig. 36 *The relationship between the density of a milk–water mixture and the descent of a drop through a solution of copper(II) sulphate*

2.4 Amylase production in germinating grains of barley

1 and 2 See Table 27 and Fig. 37.

Table 27

Times of testing for starch (min)	Age of batches (days)						
	1(G)	2(F)	3(E)	4(D)	5(C)	6(B)	7(A)
0	+	+	+	+	+	+	+
8	+	+	+	+	+	+	+
16	+	+	+	+	+	+	−
24	+	+	+	+	+	−	−
32	+	+	+	+	+	−	−
40	+	+	+	+	−	−	−
48	+	+	+	−	−	−	−
56	+	+	−	−	−	−	−
64	+	+	−	−	−	−	−
72	+	−	−	−	−	−	−
80	−	−	−	−	−	−	−

When whole barley brains were germinated, levels of amylase increased within the grains, rising on each successive day of germination. Therefore, the graph should be plotted in a way

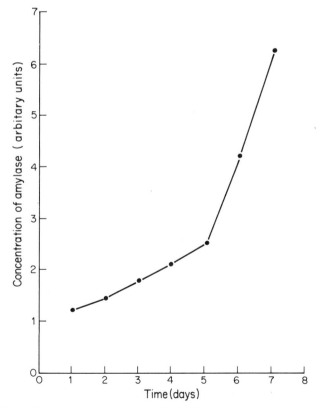

Fig. 37 *Amylase production in germinating grains of barley. Note: amylase activity in dry grains (day 0) was not determined*

that shows an increase in amylase activity. Arbitrary units have to be used. These can be expressed as reciprocals, $(1/t) \times 100$ of the time taken for the amylase from each batch of grains to effect complete digestion of the starch solution. Hence, the level of amylase activity in batch A can be expressed as $(1/16) \times 100 = 6.25$. Other values are given in Table 28.

Table 28

Batch of grains	Amylase acitivity (arbitrary units)
A	6.25
B	4.2
C	2.5
D	2.1
E	1.8
F	1.4
G	1.25

3 See Fig. 38. Times were recorded for the filtrates from both the embryo-halves and endosperm-halves of the grains to effect complete digestion of the starch solution. These were: embryo-halves—40 minutes, endosperm-halves—72 minutes. Histograms were plotted from reciprocals, $(1/t) \times 100$, of these times, to indicate a higher level of amylase activity in the embryo-halves than in the endosperm-halves.

The rise in amylase levels observed in whole grains occurs as a result of enzyme induction at the aleurone layer of the endosperm. Gibberellic acid, transported by water from the embryo to the aleurone layer, induces the synthesis of α-amylase, which, in turn, begins the hydrolysis of starch stored within the endosperm of the grain.

In half-grains, where the embryo and endosperm are separated from one another, amylase synthesis is impaired, but nevertheless occurs in embryo-halves, where sufficient of the endosperm remains in contact with the embryo to permit enzyme synthesis to take place.

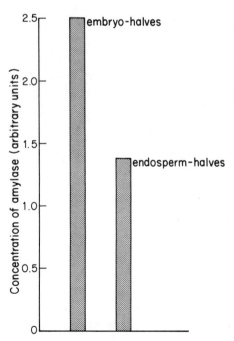

Fig. 38 *Amylase production in immersed embryo-halves and endosperm-halves of barley grains*

2.5 Variations in the rate of salivary amylase activity

4 See Fig. 39.

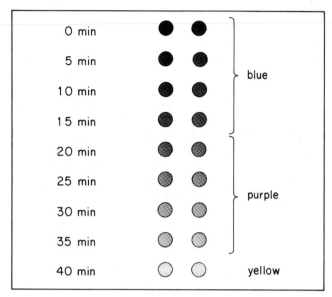

Fig. 39 *Colour changes occurring in the paper discs at five-minute intervals*

5 See Table 29 and Fig. 40. No of students in class = 25

Table 29

Time for completion of starch digestion (min)	No. students completing digestion within each time limit
0	0
5	0
10	0
15	0
20	0
25	0
30	0
35	1
40	6
45	9
50	5
55	3
60	1

The distribution of variation in amylase activity, measured in a class of 25 students, was found to be continuous. Only a small number of students possessed very active or relatively inactive amylase, while the majority possessed amylase that was similar and comparable in terms of the rate at which it hydrolysed unit masses of starch. Whilst the reasons for this variation in amylase activity are not fully understood it may be that they are related to:

(i) genetic factors, which determine amounts and activity of the amylase produced,

(ii) diet, including the amount of starchy and non-starchy foods normally consumed, or

(iii) behaviour especially smoking, which may reduce amylase activity, or exercise, which may increase it.

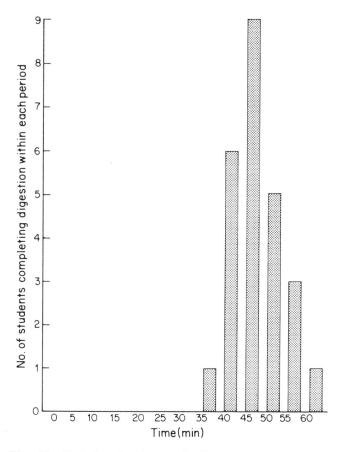

Fig. 40 *Variations in the rate of salivary amylase activity*

6 Amylase hydrolyses starch, via dextrins, to maltose:

$$\text{Starch} \longrightarrow \text{Dextrins} \longrightarrow \text{Maltose}$$

Whilst starch gives a blue-black coloration with iodine solution, and dextrins give a purple-blue coloration, maltose effects no colour change in the iodine solution. Hence, iodine may be used to indicate both the stages in the reaction and the end point of the reaction.

Thin starch-impregnated paper, such as that used in this experiment, consists of a bi-layer of starch overlying a cellulose matrix. Provided that discs of uniform size are used, amounts of starch used for each test should be uniform. The method has several advantages over the use of iodine solution mixed with a suspension of starch on a white tile, namely that:

(i) it is difficult, when applying reactants from droppers, to ensure that all drops, both of iodine solution and suspension of starch, are of uniform size;

(ii) when starch is used in suspension, the starch grains sediment under the influence of gravity, so that the density of starch grains may vary throughout the mixture.

The use of starch-impregnated paper removes both of these difficulties.

2.6 The effects of α-amylase and β-amylase on the digestion of starch

2 See Table 30 and Fig. 41.

Table 30

Time (h)	Gain in mass of tubes (g)		
	Tube 1 α-amylase	Tube 2 β-amylase	Tube 3 α-amylase + β-amylase
0	0	0	0
0.5	0.025	0.05	0.1
1.0	0.05	0.2	0.4
1.5	0.075	0.6	1.05
2.0	0.1	1.05	1.625
2.5	0.15	1.375	2.30
3.0	0.2	1.6	2.75

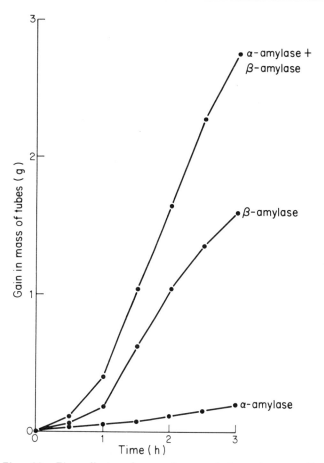

Fig. 41 *The effects of α-amylase and β-amylase on the digestion of starch*

3 (i) Amylase catalyses the hydrolysis of starch via a series of compounds:

$$\text{Starch} \longrightarrow \text{Dextrins} \longrightarrow \text{Maltose}$$

Starch itself is osmotically inactive. Therefore, if starch alone is present in the cellophane sac, no water will be drawn in by osmosis and there will be no gain in the

105

weight of the contents of the sac. Similarly dextrins (e.g. amylase) are osmotically inactive. If dextrins were formed from the starch there would again be little, if any, gain in the mass of the sac. Maltose, however, is osmotically active and water will enter the sac if maltose is present. The rate at which water enters provides an index of maltose concentration, the more maltose formed, the more rapidly will water enter the sac, causing the mass to increase.

(ii) α-amylase acts on amylose to split the more central linkages, forming a mixture of smaller dextrins but very little maltose. β-amylase splits maltose molecules from both ends of the amylose molecule, hydrolysing it eventually to maltose. When both enzymes act in combination, both the central linkages and the terminal units of amylose are attacked, resulting in the more rapid hydrolysis of the starch to maltose. This is reflected in the greater gain in mass shown by Tube 3.

2.7 The hydrolysis of sucrose by invertase

2 See Table 31 and Fig. 42.

Table 31

Time after addition of invertase (min)	Time for decolourisation of manganate(VII) (s)	Equivalent concentration of glucose (g/100 cm³)
1	420	2.5
5	150	6.5
10	130	7.5
15	120	8.0
20	115	9.0
25	115	9.0
30	115	9.0
60	115	9.0
90	115	9.0

3 Results indicate that the enzyme invertase causes rapid hydrolysis of sucrose. After 25 minutes, however, the reaction appears to have reached a point of equilibrium, at which levels of reducing sugars remain constant. As the actual levels of reducing sugars remain below those levels predicted from theoretical considerations, it appears likely that the mixture, after 30 minutes, consists of a mixture of sucrose, fructose and glucose.

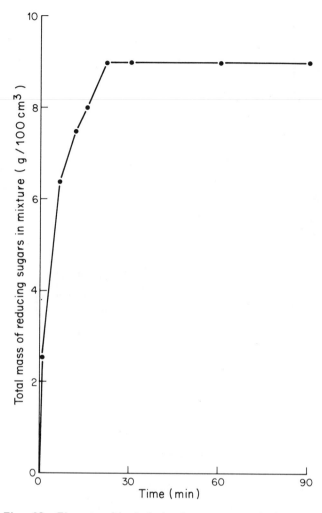

Fig. 42 *The rate of hydrolysis of sucrose to reducing sugars*

2.8 To determine the rate at which milk is coagulated by rennin at room temperature

3 See Table 32 and Fig. 43.

Table 32

Period of incubation (min)	Time taken for drainage of syringe (s)
0	8
10	8
20	9
30	15
40	72
50	87
60	210
70	600
80	900
90	no drainage

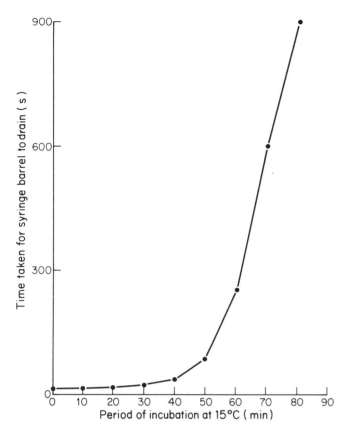

Fig. 43 *The rate of coagulation of milk at 15°C. Gelation occurs after 90 minutes*

4 As the reaction proceeds the milk becomes more viscous and therefore drains more slowly, under gravity, from the barrel of the syringe. When gelation is complete the milk is retained, as a solid mass, within the container. Results show that the increase in viscosity is not uniform throughout the period of the reaction. Initially, the increase in viscosity is slow and it is not until 40 minutes after the enzyme and milk are mixed that there is a marked increase in viscosity. Once this point has been reached there is a rapid increase in viscosity, until the milk is in the form of a solid gel after 90 minutes. One problem that may be encountered is blockage of the nozzle of the syringe by small pieces of curd, but generally solid blocks are not formed providing the experiment is carried out within the temperature range 15–18°C.

5 (i) The clotting of fresh blood, under the influence of enzymes such as thrombin.
  (ii) The digestion of gelatine by proteolytic enzymes, such as pepsin or trypsin, or the breakdown of pectin by pectolases.

2.9 The effect of temperature on the rate of an enzyme-catalysed reaction

2 Almost immediately after the addition of cold hydrochloric acid to the suspension of Marvel milk, the turbid suspension of milk particles cleared, possibly as a result of the particles dissolving in the acid solution. If this interpretation is correct, then the most probable explanation is that the acid, in the absence of an enzyme, is capable of hydrolysing the milk protein to soluble breakdown products.

4 See Table 33.

Table 33

Temperature of reaction (°C)	Time for completion of reaction (min)
14	16
24	8
34	4
44	2
54	3.5
64	not completed

5 See Fig. 44. Within the temperature range 0–40°C, each rise in temperature of 10°C causes an approximate doubling of the rate of enzyme activity, known as the Q_{10} of the enzyme. At temperatures above 40°C, however, the enzyme begins to undergo denaturation, so that the rate is slowed, and ceases completely when all molecules of the enzyme have been denatured.

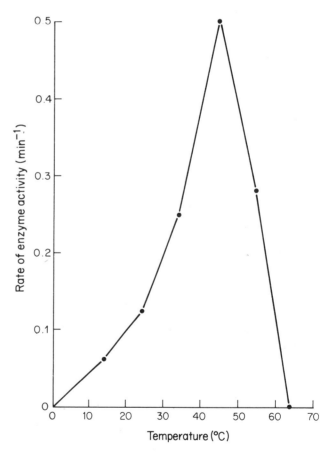

Fig. 44 *The relationship between temperature and the rate of enzyme activity*

6 It would be preferable to control pH by the use of buffer solutions. Additionally, the trypsin solution should be left at 64°C only long enough to have reached this temperature, but not for as long as five minutes, which would have caused further denaturation and led to an erroneous result.

2.10 To determine the effect of temperature on the inactivation of trypsin

3 See Table 34.

Table 34

Duration of heat treatment (min)	Time for completion of reaction (min)	
	60° C	80° C
0	4.0	4.0
5	5.5	29.0
10	7.0	–
15	10.0	–
20	13.0	–
25	16.0	–
30	21.0	–

4 See Fig. 45.

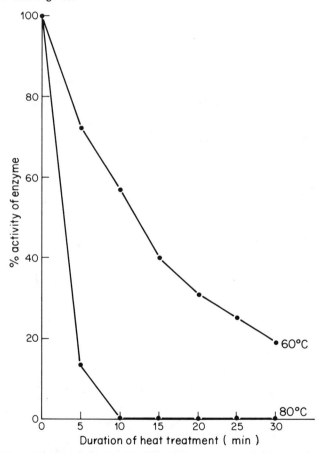

Fig. 45 *The effect of temperature on the inactivation of trypsin*

5 Relatively high environmental temperatures of 60° C and 80° C cause progressive denaturation of the enzyme, the degree of inactivation being dependent on, and directly related to, the duration of the heat treatment. At a temperature of 60° C denaturation occurs relatively slowly, so that the enzyme is not completely denatured at the end of a 30 minute period of heat treatment. There is therefore a need, at this temperature, to continue with the heat treatment in order to determine the time taken for complete denaturation of the enzyme. Conversely, at 80° C the enzyme is completely denatured after 10 minutes and there is therefore a need to take more readings over this initial period of treatment, withdrawing samples of the enzyme at, say, intervals of two minutes.

It would be instructive to broaden the scope of the experiment by setting up water-baths at 30, 40, 50, 60, 70, 80, 90 and 100 ° C and to determine the rate of enzyme inactivation at each of these temperatures, continuing the treatment until the enzyme was completely inactivated at each temperature of incubation.

2.11 To determine the optimal pH for the activity of two enzymes

2 See Table 35 and Fig. 46.

Table 35

pH	Concentration of ammonia (p.p.m.)	
	15 min	60 min
4.0	0	10
6.4	10	60
6.8	30	80
7.4	60	100
8.0	30	60
9.0	10	30

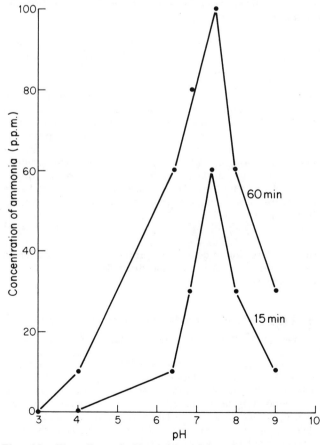

Fig. 46 *The effect of pH on the activity of urease*

The optimal pH for urease derived from soya bean is in the range 7.2–7.9. This enzyme, however, shows activity over a wide range of pH values, a feature that may be related to the role of the enzyme in breaking down substances which are potentially harmful to organisms in which the enzyme occurs. The optimal pH for the fermentation of glucose by yeast lies within the range 6.0–6.5. Zymase, however, is not a single enzyme, but a complex of enzymes each with optimum activity at a different pH value. Fermentation therefore occurs over a wide range of pH values.

5 See Table 36 and Fig. 47.

Table 36

pH	Volume of mixture displaced or collected (cm^3)
4.0	1.5
6.4	4.4
6.8	3.8
7.4	3.0
8.0	2.5
9.0	2.3

2.12 The effect of pH on the activity of two proteolytic enzymes

4 See Tables 37 and 38 and Fig. 48.

Table 37 *Results for pepsin (labelled A).*

Time (min)	Area of gelatine (mm^2)				
	pH 4.0	6.4	7.4	8.0	9.0
0	200	200	200	200	200
15	212	214	216	218	220
30	166	192	210	218	220
45	104	186	209	218	220
60	52	145	205	218	220
75	15	121	200	219	220
90	0	100	196	218	220

Table 38 *Results for trypsin (labelled B).*

Time (min)	Area of gelatine (mm^2)				
	pH 4.0	6.4	7.4	8.0	9.0
0	200	200	200	200	200
15	210	212	212	210	210
30	210	212	200	174	164
45	210	206	168	130	108
60	210	200	122	108	43
75	210	194	108	82	8
90	210	192	90	48	0

Of the two enzymes, A appears to be more active than B (assuming that both are supplied in the same concentration). Enzyme A shows optimum activity in acid pH, while enzyme B has an optimum within the alkaline range. These differences suggest that both enzymes are from the alimentary canal of an animal.

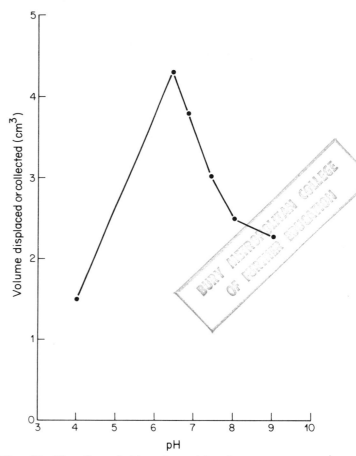

Fig. 47 *The effect of pH on the activity of zymase*

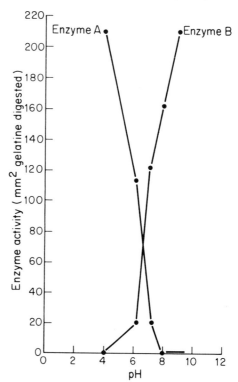

Fig. 48 *The effect of pH on the activity of two proteolytic enzymes*

5 (i) The strips of gelatine were not of uniform thickness.
(ii) The experiment was not controlled. It is essential to discover if the gelatine strips will dissolve in water only, acids or alkalis, without the addition of an enzyme. Additionally, it is desirable to know if the enzyme, without the buffer, has any effect on the substrate.
(iii) Temperature was not controlled.
(iv) An insufficiently wide range of pH buffers was used.
(v) It would have been preferable to have used two or more strips of gelatine per dish.
(vi) It would have been preferable to have mixed the enzyme with the buffer before the mixture was added to the gelatine.

2.13 The effect of pH on phosphatase activity in the testis and epididymis of a male rat

5 See Table 39 and Fig. 49.

Table 39

pH of buffer	Colour intensity of product (arbitrary units)
4.0	7.0
6.4	3.0
7.4	2.0
8.4	8.0
9.4	9.0

The investigation indicates that there is phosphatase activity over a wide range of pH. As this is unlikely to result from the activity of a single enzyme, it seems likely that there are at least two enzymes present in the tissue extract, one an acid phosphatase and the other an alkaline phosphatase, with the alkaline phosphatase either present in larger amounts, or in a more active form that the acid phosphatase.

6 (i) It would have been preferable to have made separate investigations of phosphatase activity in the testis and epididymis, rather than to macerate these two tissues together.
(ii) A wider range of pH buffers should have been used, both below 4.0 and above 9.4. Additionally, a greater number of buffers should have been used e.g. pH 2.5, 3.0, 3.5 etc. to 12.0.
(iii) A colorimeter should have been used to measure the absorbance of each tissue/enzyme substrate mixture after the addition of a standard volume of sodium carbonate solution.

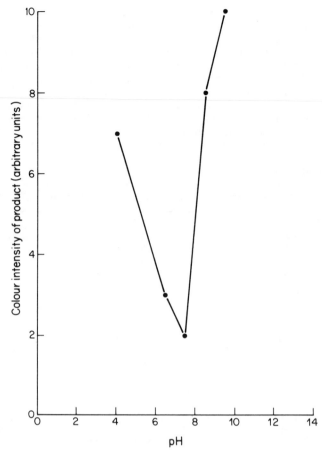

Fig. 49 *The relationship between pH and phosphatase activity in the testis and epididymis of a rat*

2.14 The effects of pH and lysozyme on the rate of sedimentation of an aqueous suspension of yeast

2 See Table 40.

Table 40

pH	Depth of sediment (cm)	Depth of yeast-free column (cm)
4.0	0	6.5
6.4	0.8	0.5
7.0	0.8	0.5
8.0	0.8	0.5
9.0	0.8	0.5

Over the pH range 6.4–9.0 yeast sediments rapidly from the suspension, probably as aggregates of cells, that collect at the bottom of the vessel. Some yeast cells, however, remain in suspension, so that only a small yeast-free zone has formed at the top of the column at the end of a 30 minute period.

At pH 4.0 the pattern of sedimentation is different. Few, if any, yeast cells have sedimented at the bottom of the vessel. Most of the yeast remains in a homogeneous suspension, which sediments slowly under the influence of gravity, leaving a less turbid, more distinct zone above the homogeneous mass of sedimenting cells.

3 See Table 41.

Table 41

pH	Depth of sediment (cm)	Depth of yeast-free column (cm)
4.0	0	7.5
6.4	0	4.0
7.0	0.8	0.5
8.0	0	3.7
9.0	0	4.7

Lysozyme has little, if any, effect on the pattern or rate of sedimentation at pH 7.0, but at pH 4.0–6.4, and again at pH 8.0–9.0, the pattern and rate of sedimentation are modified by the presence of the enzyme. Rapid sedimentation of large aggregates of cells is replaced by slower sedimentation of a homogeneous mass of cells. The rate at which the homogenate undergoes sedimentation is directly related to pH, being most rapid at the two ends of the pH range, and slowest at the mid-point (pH 7.0).

The effect of lysozyme demonstrated in this experiment may result from the ability of molecules of this enzyme to act as polyanions, affecting the distribution of yeast cells in suspension, by either attracting them or repelling them, depending on the charge present on the surface of the cell wall.

4 See Fig. 50.

(i) Depth of sediment at base of vessel

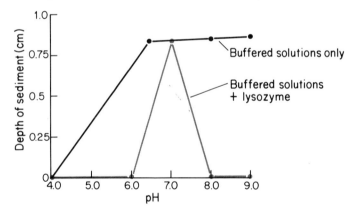

(ii) Depth of yeast-free zone

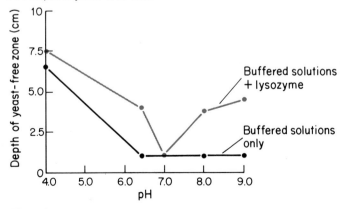

Fig. 50 *The effect of pH and lysozyme on the rate of sedimentation of yeast*

5 (i) Investigate the effects of pH, in the absence of lysozyme, on rates of sedimentation of yeast cells over the pH range 1–6.
  (ii) Determine the optimum pH for lysozyme activity, measured by the enzyme's ability to hydrolyse mucopolysaccharides in the cell wall of yeasts.

2.15 The effect of ascorbic acid (vitamin C) on levels of diphenol oxidase activity

1 When cut segments of apple and potato are exposed to air, the white flesh slowly browns, the browning generally occurring more rapidly in the potato than in the apple. The browning process has commercial importance as it may cause spoilage of apple segments prepared for fruit salads, drying etc., and potatoes prepared for chipping or frying.
2 The apple contained approximately 5 mg ascorbic acid/100 g.
3 The potato contained approximately 10 mg ascorbic acid/100 g.
4 Both the apple and potato contain diphenol oxidase, which as soon as the tissues are cut or bruised, oxidises a naturally-occurring phenol, also present in the tissues. If a solution of catechol is added to macerated tissue, the browning is accelerated and the final colour is darker. This effect occurs because diphenol oxidase can use catechol as an artificial substrate.

Ascorbic acid, a powerful reducing agent, inhibits the oxidation of phenols by diphenol oxidase. As a result of this reduction, the browning of the tissues is either slowed or inhibited.

Hydrogen peroxide, a powerful oxidising agent, accelerates the oxidation of phenols by the enzyme. In the presence of oxygen, or an oxidising agent, the rate at which browning occurs is accelerated.
5 If cut segments of apple or potato are exposed to air, they soon become spoiled by browning. This colour change, which results from the action of diphenol oxidase on phenols, can be slowed or inhibited by reducing agents, or accelerated by oxidising agents. As the process of browning is inhibited by ascorbic acid, which is present both in apples and potatoes, breeders might attempt to produce cultivars with a higher vitamin C content, or those concerned with the preservation of food might add vitamin C to apple segments or peeled potatoes to prevent deterioration.

2.16 The effect of (i) concentration of enzyme and (ii) concentration of enzyme substrate on the rate of an enzyme-catalysed reaction

3 Fig. 51 shows that most of the gas evolved from the reaction passes into the collecting vessel within one minute of mixing the two reactants. Thereafter, the rate slows and very little more oxygen is evolved between the fourth and fifth minutes. The pattern of oxygen evolution indicates that the reaction is extremely rapid.

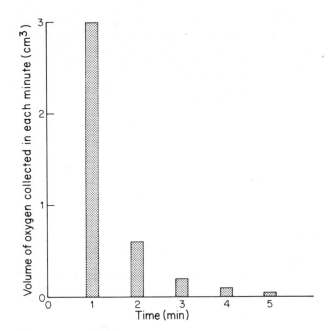

Fig. 51 *Evolution of oxygen from a reaction between hydrogen peroxide and yeast*

4 See Table 42.

Table 42

Volume yeast suspension (cm³)	Volume oxygen evolved (cm³)
5	4.8
10	5.0
15	4.9
20	5.1
25	5.1
30	5.0

5 See Fig. 52. Generally, if enzyme concentration is increased and the amount of substrate remains constant, there is a linear relationship between the concentration of the enzyme and the rate of the reaction. In this case, however, the graph shows that the volume of gas is constant, irrespective of the amount of enzyme present. Hence, it seems probable that enzyme is present in excess, and that the reaction is limited by the amount of substrate used.

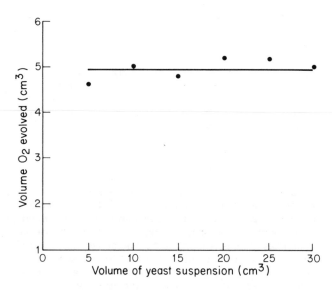

Fig. 52 *The relationship between concentration of enzyme and rate of reaction*

6 Set up apparatus as illustrated in Fig. 8. Fill a 10 cm³ plastic syringe with hydrogen peroxide solution, and connect the syringe to the needle. Using 5 cm³ yeast suspension in the boiling tube for each test, measure and record the volume of oxygen evolved following the addition of 1, 2, 3, 4, and 5 cm³ hydrogen peroxide solution to the yeast. Take each reading after a period of 5 minutes, tabulate the results (Table 43), and plot a graph (Fig. 53).

Table 43

Volume hydrogen peroxide solution (cm³)	Volume oxygen evolved (cm³)
1	4.9
2	9.9
3	15.1
4	20.0
5	24.8 (read after 2.5 minutes and corrected)

If substrate concentration is increased, and the amount of enzyme remains constant, there is an initial linear relationship between the concentration of substrate and the rate of the reaction. If, however, excess substrate is added, the reaction is limited by the amount of enzyme present, and the reaction proceeds at a constant, maximum rate without any further increase. The effect of adding increased volumes of hydrogen peroxide to the yeast is to increase the amount of oxygen evolved, the increase being linear over the range of concentrations used. Hence, in this experiment, enzyme substrate is not present in excess.

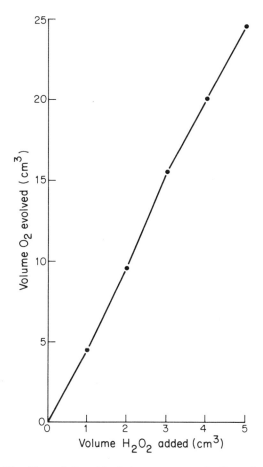

Fig. 53 *The relationship between concentration of enzyme substrate and the rate of reaction*

2.17 Nitrate reductase and nitrite reductase activity in seedlings of dwarf French bean

2 See Table 44 and Fig. 54.

Table 44

Time (min)	Concentration of NO_2^- (p.p.m.)		
	Leaves	Stems	Roots
0	0	0	0
15	0	0	1
30	1	0	5
45	5	0	10
60	10	1	25
75	10	1	25
90	25	1	50

Nitrate reductase is present both in the roots and in the leaves of *Phaseolus vulgaris*. Generally, larger quantities of the enzyme occur in the roots than in the leaves, but these amounts are variable, dependent upon the nitrogen status of the soil in which the plants have grown. Stems generally show some nitrate reductase activity, which increases with time, as the enzyme is inducible, synthesised in response to the presence of nitrate.

4 See Table 45 and Fig. 55.

Table 45

Time (min)	Concentration of NH_4^+ (p.p.m.)		
	Leaves	Stem	Root
0	0	0	0
15	0	0	0
30	trace	0	0
45	10	0	0
60	10	0	0
75	30	0	0
90	30	0	0

Nitrite reductase occurs only in leaves, where it is believed to be associated with chloroplasts.

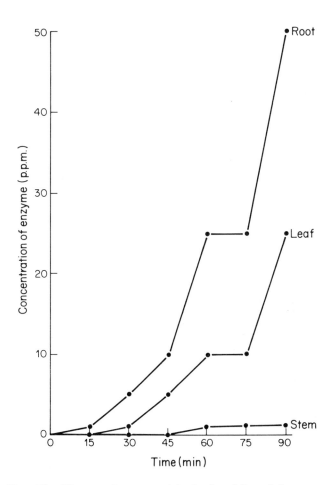

Fig. 54 *Nitrate reductase activity in dwarf French bean*

113

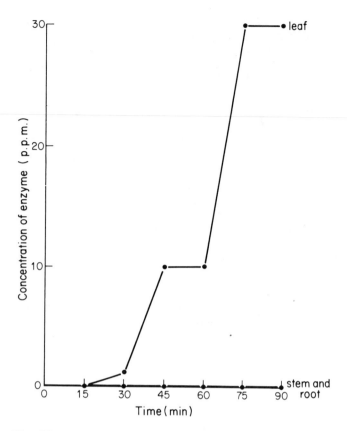

Fig. 55 *Nitrite reductase activity in dwarf French bean*

5 Plants absorb nitrate (NO_3^-) from the soil. In order that the nitrate can be of use to the plant it must first be converted to ammonia. This is a two-step process, catalysed by nitrate reductase and nitrite reductase:

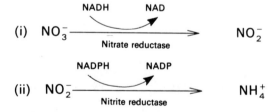

The ammonia is then assimilated to glutamate in reactions catalysed by glutamate dehydrogenase, glutamine synthetase and glutamate synthase.

The test provides a useful index of the nitrogen-nutrition status of the plant. If the plant is receiving an adequate amount of nitrogen, levels of nitrate reductase and nitrite reductase are likely to be high. Conversely, under conditions of nitrogen deficiency, plants will produce relatively small amounts of these two enzymes. Hence, the test has an important practical use.

2.18 Enzyme induction in seeds of mung bean

2 Tests applied to Tube 1 proved negative over a period of 24 hours. Therefore glucose is not leached from the seeds as a result of soaking, nor is nitrite derived from any stored material present in the dried seed.
The Merckoquant test for nitrite proved positive in Tube 2, the concentration of nitrite increasing with time (Table 46 and Fig. 56).

Table 46

Time (h)	Nitrite in solution (p.p.m.)
0	0
4	0
8	5
12	15
16	30
20	45
24	50

Additionaly, the Clinistix test proved positive in Tube 3, concentrations of glucose increasing throughout the duration of the experiment (Table 47 and Fig. 57).

Table 47

Time (h)	Glucose in solution (g/100cm³)
0	0
1	0
2	0
3	0.1
4	0.25
5	0.6
6	0.75

3 The positive test for nitrite in Tube 2 and for glucose in Tube 3 indicates that enzymes capable of breaking down the two substrates have been produced by the seeds in response to the presence of enzyme substrate supplied to the seeds. Hence, production of the enzymes nitrate reductase and invertase afford additional examples of enzyme induction, first demonstrated by Jacob and Monod, who showed that the bacterium *E. coli* would synthesise the enzymes β-galactosidase and permease only when the enzyme substrate, lactose, was present in the growth medium, but not at other times.

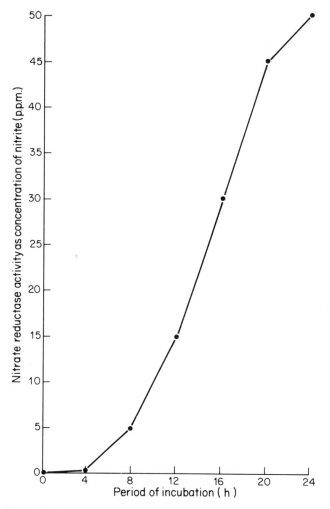

Fig. 56 *The induction of nitrate reductase activity in mung beans*

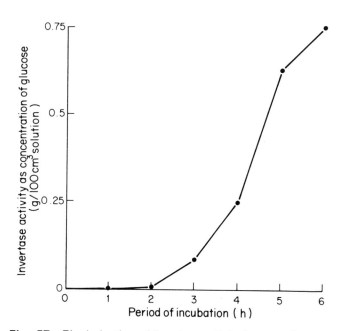

Fig. 57 *The induction of invertase activity in mung beans*

2.19 Staining techniques for the identification of fats and proteins in maize grains

(i) See Fig. 58. Fats are contained chiefly in the embryo region of the grain, with smaller amounts evenly distributed throughout the endosperm.
(ii) See Fig. 59. Proteins are present throughout the grain. The largest concentrations of protein, however, occur in the embryo region and in the outer region of the endosperm, together with the aleurone layer.

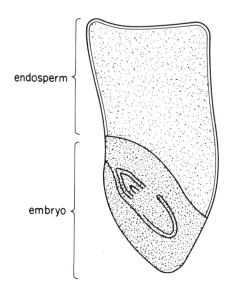

Fig. 58 *A maize grain stained with Sudan blue*

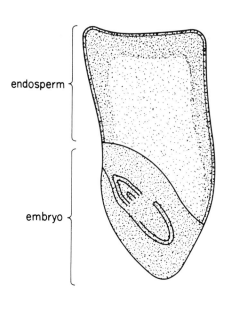

Fig. 59 *A maize grain stained with Ponceau S*

3.1 Seed structure and germination in mung bean and maize

1 See Fig. 60.

Mung bean

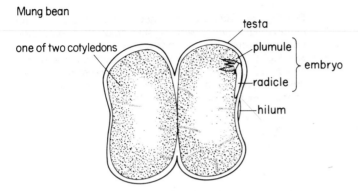

Maize

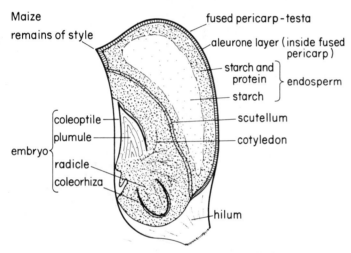

Fig. 60 *The structure and staining pattern of seeds*

(i) Mung beans are true seeds, whereas maize grains are one-seeded fruits.
(ii) Mung bean seeds contain two cotyledons; maize grains contain a single cotyledon.
(iii) Mung bean seeds are exalbuminous (without an endosperm); maize grains are albuminous (with an endosperm).

2 See Table 48.

Stain X, an iodine solution, stains starch blue and dextrins purple. Dormant seeds store starch, which, as the seed germinates, is converted to dextrins and other degradation products. These pass from the region of storage into the embryo of the seed. Dextrins passing through the cotyledon of a maize grain to the embryo can be visualised by application of this stain.

Stain Y, a solution of silver nitrate, stains areas of the seed that contain reducing sugars, notably glucose. Metallic silver is deposited in regions containing glucose. These regions are chiefly those with a high metabolic rate, such as the apical meristems of the plumule and radicle.

Stain Z, tetrazolium salt, is a vital stain, reduced by dehydrogenases to an intensely red formazan. Dehydrogenase activity is associated with mitochondria. Hence, the stain locates and visualises active mitochondria, which occur in all living cells, but are most abundant in cells

Table 48

Stain	Bean seed	Maize grain
X (Iodine)	Blue colour, indicative of starch, throughout cotyledons.	Starch present in endosperm. Purple colour, indicative of dextrins, present in cotyledon.
Y (Silver nitrate)	Metallic silver, indicative of reducing sugars, present in embryo, cotyledon stalks and in part of cotyledon adjacent to embryo.	Reducing sugars present in embryo, especially in apical meristems of the plumule and radicle.
Z (Tetrazolium salt)	Red coloration, indicative of mitochondrial activity, most strongly developed in embryo but also present throughout cotyledons.	Active mitochondria present in aleurone layer of endosperm and in the embryo especially apical meristems of the plumule and radicle.

with a high metabolic rate, such as those of apical meristems.

3 See Fig. 61.
  (i) Mung bean is a dicotyledon, whereas maize is a monocotyledon.
  (ii) Germination in mung bean is epigeal; germination in maize is hypogeal.
  (iii) The first foliage leaves of mung bean are net-veined, whereas those of maize are parallel-veined.

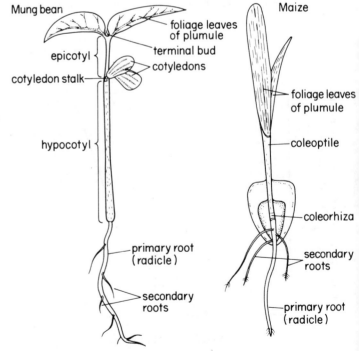

Fig. 61 *Seedlings of mung bean and maize (twice life size)*

3.2 Tests for the viability of pea and bean seeds

1 Viable runner bean seeds exude glucose when soaked in water. This exudate gives a positive test with Clinistix. Viable pea seeds, however, exude sucrose, which does not react with Clinistix. The addition of invertase effects hydrolysis of the sucrose to a mixture of glucose and fructose. On applying the second test with Clinistix, the glucose fraction gives a positive result.

2 See Table 49 and Fig. 62.

Table 49

Reaction with Clinistix	No. seeds showing reaction	
	Pea	Bean
Negative	1	1
Light	2	1
Medium	5	8
Dark	12	10

3. See Fig. 63.

4 In viable seeds certain enzymes are present, which lose their potency, either partly or completely, as the seed ages. Viable seeds of pea and bean exude sugars, and contain dehydrogenases which reduce tetrazolium salt to an intensely red pigment called a formazan. Viable seeds exude the largest amounts of sugars and contain the most active dehydrogenases. Therefore, either of these tests may be used to assess the viability of a sample of seed, without the need for germination. Whilst the Clinistix test may be used to assess the viability of pea and bean seeds, it is not applicable to all seeds, as many germinating seeds do not exude sugars. The tetrazolium test is probably the most useful, and is applicable to all types of seed. This compound is reduced to a red pigment in those regions of the seed in which mitochondria are active. Seeds which show no reaction are dead. Seeds which fail to stain in the embryo region are also dead, even if the cotyledons contain active mitochondria. Staining of a part of the embryo, such as the plumule or radicle, can only be interpreted by those who are expert in the study of particular species.

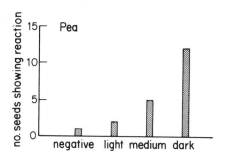

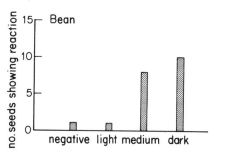

Fig. 62 *Reactions of pea and bean seeds to Clinistix strips*

(i) Negative reaction with Clinistix

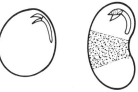

No staining of the embryo; little, or no staining of the cotyledons

(ii) Light reading with Clinistix

Certain parts of the embryo are unstained. The central part of cotyledons is unstained

(iii) Medium reading with Clinistix

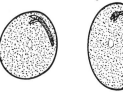

The entire embryo is stained, but the central region of the cotyledons is unstained

(iv) Dark reading with Clinistix

Both the embryo and cotyledons are fully stained

Fig. 63 *The appearance of seeds treated with tetrazolium salt*

3.3 A morphological and biochemical study of normal and etiolated pea seedlings

1 See Fig. 64.

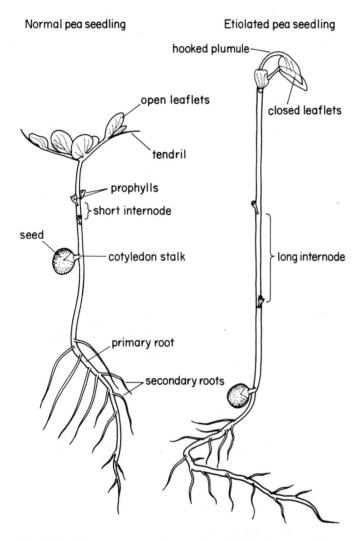

Fig. 64 *The appearance of normal and etiolated pea seedlings*

2 See Table 50.

Table 50

Normal seedlings	Etiolated seedlings
Leaves unfolded, erect	Leaves folded, sometimes hooked
Leaves green	Leaves yellow
Plumule unfolded	Plumule hooked
Short internodes	Long internodes
Robust, broad stem	Thin, fragile stem

4 See Fig. 65. Leaves taken from normal pea seedlings contained a carotene (orange), two chlorophylls (green) and at least three, possibly four, xanthophylls (yellow). Leaves taken from etiolated seedlings probably contained the same pigments, but in smaller amounts. One of the chlorophylls and one of the xanthophylls were present in very small quantities.

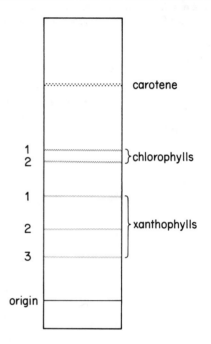

Fig. 65 *Photosynthetic pigments in a normal pea seedling*

3.4 An investigation of the Hill reaction

2 See Table 51

Table 51

Distance of chloroplasts from light source (cm)	Time for reduction of manganate(VII) solution (min)
5	8
10	14
20	19
Darkness	50

3 See Table 52.

Table 52

Distance of chloroplasts from light source (cm)	Concentration of Fe^{2+} ions (p.p.m.)
5	50
10	25
20	3
Darkness	Nil

4 Hill demonstrated that isolated chloroplasts were able to effect reduction of a wide range of substrates, including Fe^{3+} to Fe^{2+}. Isolated chloroplasts, placed into an isotonic solution, continue to generate $NADPH_2$, providing they are illuminated. Amounts of $NADPH_2$ generated are dependent on: (i) the intensity of illumination, and (ii) the duration of illumination.

Reduction of potassium manganate(VII) solution in an acid medium provides a convenient method for demonstrating the production of a reducing agent in the Hill reaction. Reduction of Fe^{3+} to Fe^{2+}, measured semi-quantitatively by the use of reagent sticks, provides a simple, convenient and highly sensitive method for determining the reducing activity of chloroplast isolates under different environmental conditions.

3.5 Measuring the rate of photosynthesis in a brown seaweed

1 See Table 53.

Table 53

Distance of beaker from lamp (cm)	Time taken for 3 discs to lift from surface (min)
5	23
10	36
15	52
20	88

2 See Table 54. The purpose of taking a reading without the alga being present, was to determine the rate at which water in the bottle expands, following heating by the illuminated lamp. In darkness, the alga was respiring, releasing CO_2 into the surrounding water. The total volume of CO_2 respired $(0.4521 - 0.0048)$ was therefore $0.4473 \, cm^3$ per hour. In light, oxygen was evolved as a product of photosynthesis. The total volume of oxygen evolved was therefore $(5.1245 - 0.4569) = 4.6676 \, cm^3$ per hour.

Table 54

Experimental condition	Distance travelled by meniscus (cm)	Vol. gas evolved per hour (cm^3)
Alga absent	0.1	0.0048
Alga present (darkness)	0.3	0.4521
Alga present (light)	3.4	5.1245

3 See Table 55. Before these figures could be plotted as a graph it would be necessary to:
(i) deduct appropriate figures for respiration and the expansion of water;
(ii) convert figures to cm^3 of oxygen produced per hour;
(iii) plot light intensity as a reciprocal of distance, e.g. $(1/d^2) \times 1000$, where d = distance (cm) of apparatus from lamp. This conversion is necessary as light intensity at a distance d cm from the lamp is proportional to $1/d^2$ (inverse square law).

Table 55

Distance of apparatus from lamp (cm)	Distance travelled by meniscus (cm)
5	3.4
10	1.4
15	0.9
20	0.5
25	0.1

4 *Method A* Discs cut from fronds of *Fucus* are more dense than sea water, and therefore sink to the floor of the beaker. Following illumination, however, bubbles of oxygen are formed, which adhere to both surfaces, making the disc less dense. After sufficient gas has accumulated, particularly along the underside of each disc, the disc lifts from the floor of the beaker.

Method B Bubbles of gas released from the frond of *Fucus* rise to the top of the apparatus and displace an equivalent volume of water into the capillary tube. This volume can be measured by recording the distance travelled by the meniscus in a unit period of time.

Method B has a number of advantages over method A, notably that it is quantitative, sensitive to slight changes in the environment and accurate. Results obtained by method A may be inaccurate as: (i) the mass of individual discs may vary, and (ii) individual bubbles of oxygen may escape from the surface of the discs, thereby failing to increase the buoyancy of the disc.

5 Sea water and distilled water were mixed in different proportions (Table 56). Using the same frond of *Fucus* throughout, the apparatus was filled, in turn, with each mixture. With the apparatus positioned at 5 cm from the illuminated lamp, the distance travelled by the meniscus was recorded over a period of 5 minutes. The results obtained are shown in Table 57 and Fig. 66 on page 120.

Table 56

Vol. sea water (cm^3)	Vol. distilled water (cm^3)	% sea water in mixture
0	200	100
20	180	90
40	160	80
60	140	70
80	120	60
100	100	50
120	80	40
140	60	30
160	40	20
180	20	10
200	0	0

Table 57

% sea water	Distance travelled by meniscus (cm)	Vol. gas evolved per hour (cm³)
100	3.4	5.13
90	3.3	4.98
80	3.1	4.68
70	2.9	4.38
60	2.7	4.07
50	2.5	3.77
40	2.4	3.62
30	1.9	2.87
20	1.7	2.56
10	1.6	2.41
0	1.3	1.96

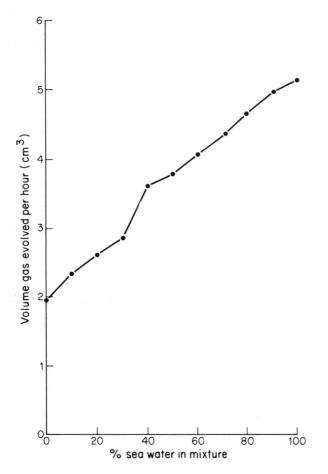

Fig. 66 *The relationship between percentage concentration of sea water and the rate of photosynthesis in a brown seaweed (Fucus serratus)*

3.6 The effect of bicarbonate ion concentrations on photosynthesis in Canadian pondweed

2 See Tables 58 and 59.

Table 58

Molarity of sodium bicarbonate solution	Volume of each component in mixture (cm³)	
	Sodium bicarbonate	Water
0.1 M	60	0
0.05 M	30	30
0.025 M	15	45
0.0166 M	10	50
0.0125 M	7.5	52.5
0.01 M	6	54

Table 59

Molarity of sodium bicarbonate solution	pH of solution
0.1 M	9.0
0.05 M	8.0
0.025 M	7.5
0.0166 M	7.4
0.0125 M	7.2
0.01 M	7.0

5 and 6 See Fig. 67.

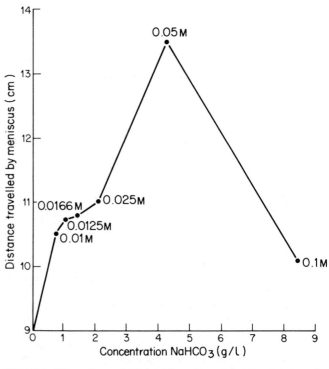

Fig. 67 *The effect of bicarbonate ion concentrations on the rates of photosynthesis in excised leafy shoots of Canadian pondweed (Elodea canadensis)*

7 (i) As only a limited number of dilutions of the 0.1 M sodium bicarbonate solution were made, it is not possible to make an accurate statement about the optimal concentration of bicarbonate ions for photosynthesis in Canadian pondweed. Even so, the optimal concentration appeared to lie close to 0.05 M, which contains 0.42 g sodium bicarbonate/100 cm^3.

(ii) This solution would contain $0.42 \times 100 \times 1000 \times 61/84 = 3050$ p.p.m. bicarbonate ions.

(iii) When the apparatus was filled with the pond water used in trials the meniscus moved a distance of 10.1 cm in 20 minutes. From the graph, this could correspond to a concentration of 0.055 g sodium bicarbonate/100 cm^3, or 0.84 g sodium carbonate/100 cm^3.

8 (i) Bicarbonate, as the HCO_3^- ion, is essential for photosynthesis in aquatic plants. Even so, there appears to be an optimal bicarbonate ion concentration for photosynthesis, which proceeds at a slower rate both above and below the optimal level. Any reduction in the rate of photosynthesis that occurs at low concentrations of bicarbonate ions may result from a rate-limiting effect, caused by a shortage of carbon atoms. At high concentrations of bicarbonate ions, however, there is a marked rise in pH, which causes a related fall in the rate of photosynthesis.

(ii) Results obtained from this investigation fail to give an accurate indication of the optimal concentration of bicarbonate ions for photosynthesis in Canadian pondweed. More extensive investigations need to be carried out, using a wider range of solutions, especially those within the range 0.025–0.05 M. Additionally, the effect of pH on oxygen production requires further investigation.

3.7 A comparison of epidermal structure and rates of transpiration in leaves of daffodil and maize

2 See Table 60 and Fig. 68.

Table 60

	Colour of indicator paper			
	Plant A (daffodil)		Plant B (maize)	
Time (min)	Upper	Lower	Upper	Lower
0	0	0	0	0
15	1	0.5	2	4
30	1.5	1	3.5	4.5
45	2	1.5	4	5
60	3	2	4.5	5
75	3.5	2	5	5
90	4	2.5	5	5

4 See Table 61.

Table 61

Species	Mean stomatal numbers/4 mm^2		Mean stomatal numbers/cm^2	
	Upper epidermis	Lower epidermis	Upper epidermis	Lower epidermis
Plant A (daffodil)	(i) 136 (ii) 146 Mean 142	67 81 74	3550	1850
Plant B (maize)	(i) 208 (ii) 200 Mean 204	264 240 252	5100	6300

5 The rate of transpiration in plant B (maize) was more rapid than in plant A (daffodil). In the daffodil, transpiration occurred more rapidly from the upper (adaxial) surface than from the lower (abaxial) surface. Generally, the rate of transpiration from any of the surfaces was directly related to the number of stomata per unit area of that surface, increasing as the density of stomata increased.

6 No. Light intensity is reduced by the paper covering, while the air beneath the paper is still, and almost dry. Therefore, other parts of the leaf are transpiring more rapidly than those covered by paper.

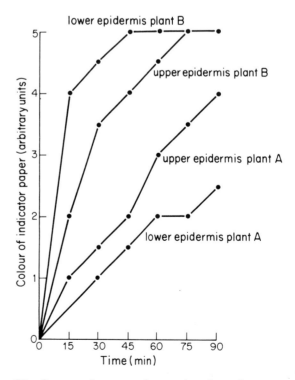

Fig. 68 *Comparative rates of water loss from the upper and lower surfaces of daffodil (plant A) and maize (plant B)*

(i) The size of each stomatal pore.
 (ii) The presence or absence of epidermal trichomes.
 (iii) The thickness of the cuticle.
 (iv) The position of guard cells relative to other epidermal cells (e.g. elevated, at the same level, or sunken).
8 (i) Air temperature.
 (ii) Light intensity.
 (iii) Relative humidity of the air.
 (iv) Wind velocity.
 (v) Availability of soil water.
 (vi) Soil temperature.
9 See Fig. 69.

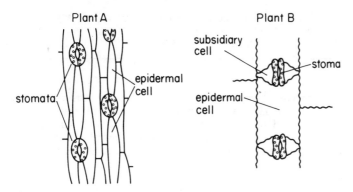

Fig. 69 *The general arrangement of stomata in the lower epidermis of daffodil (plant A) and maize (plant B)*

Plant A The epidermis consists of epidermal cells and stomata only. Epidermal cells are long and smooth-walled, arranged in longitudinal rows along the longitudinal axis of the leaf. The stomata are also arranged in parallel rows, although this arrangement is not always distinct.
Plant B The epidermis consists of epidermal cells, subsidiary cells and stomata. Epidermal cells are rectangular, approximately one-half as broad as long, with crinkled end and side walls, arranged in parallel rows. The stomata are arranged in distinct parallel rows which run along the longitudinal axis of the leaf.
10 See Fig. 70.
Plant A There are no subsidiary cells. Each guard cell is elliptical and the turgid stoma as a whole is circular in outline.
Plant B Each guard cell is subtended by a subsidiary cell. Each guard cell is shaped like a dumb-bell and the stoma as a whole is rectangular in outline.

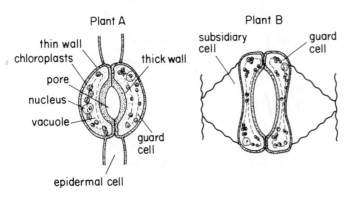

Fig. 70 *The structure of an individual stoma in daffodil (plant A) and maize (plant B)*

3.8 Measurement of water potential in inflorescence stalk cells of bluebell and dandelion

1 See Table 62 and Fig. 71.

Table 62

Molarity of sucrose	Volume molar sucrose (cm³)	Volume distilled water (cm³)
0.1M	4	36
0.2M	8	32
0.4M	16	24
0.6M	24	16
0.8M	32	8

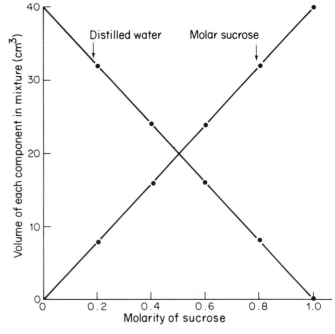

Fig. 71 *Graph for the preparation of a series of dilutions from distilled water and molar sucrose*

4 See Table 63 and Fig. 72.

Table 63

| Time (min) | Distance of split ends of stalk from central point (mm) | |
	Left-hand half	Right-hand half
0	2.5	2.5
5	9.0	4.0
10	11.0	9.0
15	12.0	11.5
20	12.5	12.5
25	14.0	14.0
30	15.5	15.5

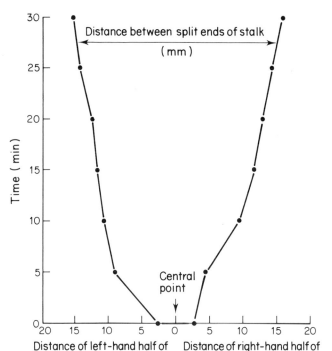

Fig. 72 *Behaviour of a partly bisected bluebell inflorescence stalk in water*

5 See Table 64 and Fig. 73.

Table 64

Molarity of sucrose	Distance between split ends of stalk (mm)		
	Stalk 1	Stalk 2	Mean value
0.0	31	29	30
0.1	18	17	17.5
0.2	9	6	7.5
0.4	2	0	1
0.6	0	0	0
0.8	0	0	0
1.0	0	0	0

A sucrose solution of concentration between 0.4 and 0.6M is equivalent to the molarity of the cell sap in the stalk cells.

6 See Fig. 74.

7 The effect of splitting the inflorescence stalk in both bluebell and dandelion was to expose parenchyma cells positioned at the centre of the stalk. At the same time tissue tension on the epidermal cells was released, so that there was some curling as soon as the stem was split. Following immersion in water, the parenchyma cells took in water by osmosis and their volume increased. Epidermal cells, on the outside of the parenchyma proved relatively impermeable to water, chiefly because they were covered by a layer of waxy cuticle. Hence, water entered these cells more slowly, if at all. The net result of this increase in the volume of parenchyma cells, while the volume of epidermal cells remained fairly constant, was an outward curvature of the split ends of the stalk, which continued to separate until the parenchyma cells were fully turgid. This movement was both more marked and more rapid in dandelion on account of the very thin layer of parenchyma that lies inside the epidermis in the inflorescence stalk of this

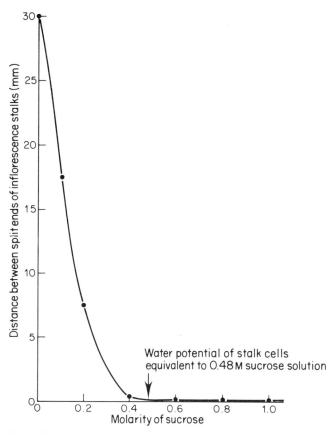

Fig. 73 *Measurement of water potential in stalk cells of bluebell*

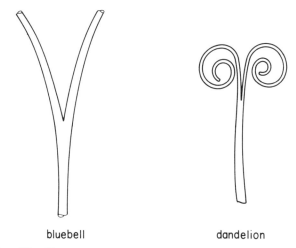

bluebell   dandelion

Fig. 74 *The appearance of lengths of bluebell and dandelion inflorescence stalks after they have stood in distilled water for 40 minutes*

plant. Bluebell, on the other hand, has a large central mass of parenchyma, into which water entered at a slower rate. Additionally, bluebell contained a number of rigid vascular strands that imposed some limitations on the degree of curvature. The combined effect was that the two split ends of the bluebell inflorescence stalk separated slowly, forming a Y shape when the parenchyma cells were fully turgid, whereas dandelion stalks were more flexible and each split end coiled to form a tight circle.

123

8 (i) The immediate curvature of dandelion inflorescence stalk, following cutting, resulted from a release of tissue tension on cells of the epidermis, causing a curvature with the epidermis on the concave surface.

(ii) The degree of curvature shown by the cut inflorescence stalks depended on differences between the volume of parenchyma cells, which were permeable to water, and the volume of epidermal cells, which were relatively impermeable. In dandelion a wax-covered epidermis overlies a thin layer of parenchyma cells. When epidermal strips were immersed in water, the parenchyma cells rapidly gained water, increasing in volume and causing the strip to coil. In bluebell parenchyma occupies the central position in the inflorescence stalk; this region is succulent and the cells are just turgid. Additional intake of water by osmosis caused slow outward curvature of the strips, a movement that was further restricted by the rigidity of the vascular bundles.

(iii) Bluebell is the most suitable material for the methods employed. Only with bluebell, in which the split ends of the inflorescence stalk remain fairly straight, is it possible to measure the distance between the split ends, which provides an index of the molarity of the solution in which the tissue is immersed. If dandelion had been used, the fact that the split ends coil into tight circles would have prevented results from being obtained.

9 More accurate results could have been obtained if the following points had been observed.

(i) The lengths of inflorescence stalk should all have been of the same diameter, and taken from the same level in the stalks of different plants.

(ii) Stalks should have been bisected mechanically, to ensure splitting into two identical halves. Splitting by hand often produces halves of unequal thickness, which curve at different rates when immersed into solutions.

(iii) The experiment should have been carried out using a wider range of dilutions of sucrose, e.g. 0.1 M, 0.2 M, 0.3 M, and so on up to 1.0M. As a result of using a small number of dilutions over the range 0.0–0.5M, it was not possible to make an accurate measurement of the water potential of the stalk cells, although this appears to be equivalent to a sucrose solution of between 0.4 and 0.5M. In order to make this determination, additional dilutions of sucrose in the range 0.4–0.5M would have to be prepared, and further results obtained.

3.9 The contribution of individual leaves to the rate of transpiration in a cut shoot of beech

2 and 3 See Table 65.

Table 65

Condition of twig	Distance travelled by meniscus in 5 minutes (cm)	Water uptake (cm³)
Intact	15.4	0.483
Two leaves greased	12.9	0.405
Two leaves removed	12.3	0.386
Four leaves removed	7.5	0.235
Six leaves removed	4.0	0.126
Eight leaves removed	0.9	0.028

4 See Fig. 75.

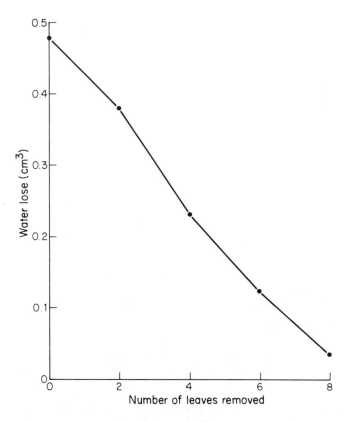

Fig. 75 *The effect of removing pairs of leaves on the rate of transpiration in a leafy shoot of beech*

5 Water loss from intact shoot = 0.483 cm³
Water loss following removal of 1st pair of leaves = 0.386 cm³
Water loss when lower surface of leaves was greased = 0.405 cm³
Therefore, water loss from 1st pair of leaves = 0.097 cm³
Water loss from upper surface of 1st pair of leaves = 0.019 cm³

$$\% \text{ water loss from upper surface} = \frac{0.019}{0.097} \times 100 = 19.58\%$$

6 Even after removal of all the leaves from the twig, water was still taken into the stem. This water leaves the stem via: (i) lenticels, (ii) the epidermis or cork and (iii) the cut ends of xylem vessels. The percentage of water lost via these routes was equal to $(0.9/15.4) \times 100 = 5.8\%$.

7 See Table 66.

Table 66

Material weighed	Mass (g)	Area (cm²)
Graph paper	3.7	100
Leaf 1	0.74	20.1
Leaf 2	1.31	35.6
Leaf 3	1.82	49.2
Leaf 4	1.32	35.8
Leaf 5	1.09	29.5
Leaf 6	1.39	37.9
Leaf 7	1.35	36.5
Leaf 8	0.75	20.3
Total leaf area		264.8

8 A total of $(0.483 - 0.28) = 0.455$ cm³ water was lost from a total leaf area of 264.8 cm² in a period of 5 minutes.

$$\text{Therefore, water loss} = \frac{0.455 \times 10\,000 \times 60}{264.8 \times 5}$$

$$= 206.2 \text{ cm}^3 \text{ m}^{-2} \text{ h}^{-1}.$$

Assuming that 19.58% of the water was lost via the upper surface, then the loss rate was 40.4 cm³ m⁻² h⁻¹ from the upper surface and 165.8 cm³ m⁻² h⁻¹ from the lower surface.

3.10 A volumetric method for investigating the water relations of succulent tissues

1 See Table 67 and Fig. 76. The molarity of solution X is 0.3 M.

Table 67

Concentration of saline solution	Original vol. water or saline (cm³)	Vol. tap water or saline after 24 hours (cm³)	
		Potato	Swede
0.0 tap water	90	84	80
0.25 M	90	91	85
0.5 M	90	95	90
0.8 M	90	95	91
1.0 M	90	95	91
solution X	90	92	86

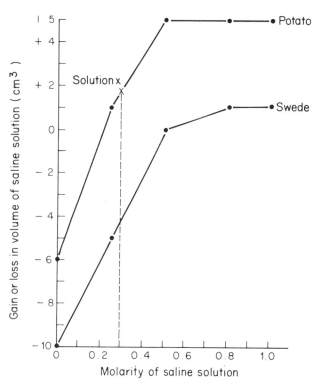

Fig. 76 *Water relations of potato and swede tissue. The osmotic pressure of the tissue is found by extrapolation to the base line*

2 The osmotic potential of potato sap is:

$$\frac{0.22}{1.00} \times 22.4 = 4.93 \text{ atmospheres.}$$

The osmotic potential of swede sap is:

$$\frac{0.5}{1.00} \times 22.4 = 11.2 \text{ atmospheres}$$

3 (i) The cups should have been covered overnight to prevent loss of water by evaporation.
   (ii) Distilled water, and not tap water, should have been used. Additionally, it would have been preferable to use 0.75 M saline rather than 0.8 M saline.
   (iii) The rods of potato should have been weighed before and after immersion in each of the solutions; this would have provided an index of gains and losses in mass, resulting from the entry of water into the tissues, or removal of water from the tissues.

4 The osmotic potential of potato sap (4.93 atmospheres) is less that one-half that of swede sap (11.2 atmospheres). These differences are probably related chiefly to the nature of the stored carbohydrates in the two tissues: potato tubers store starch, which is osmotically inactive, while swede roots store sucrose, which is osmotically active.

3.11 To determine the conditions necessary for germination of pollen grains of *Amaryllis*

1 See Table 68 and Fig. 77.

Table 68.

Solution	% of germination
A Distilled water	62
B Glucose 3 g/100 cm³	68
C Glucose 6 g/100 cm³	65
D Glucose 9 g/100 cm³	65
E Glucose 12 g/100 cm³	30
F Glucose 15 g/100 cm³	10
G Glucose 18 g/100 cm³	5
H Glucose 21 g/100 cm³	0
I Glucose 24 g/100 cm³	0
J Sucrose 3 g/100 cm³	68
K Sucrose 6 g/100 cm³	68
L Sucrose 9 g/100 cm³	68
M Sucrose 12 g/100 cm³	67
N Sucrose 15 g/100 cm³	65
O Sucrose 18 g/100 cm³	66
P Sucrose 21 g/100 cm³	45
Q Sucrose 24 g/100 cm³	30

2 See Fig. 78.

3 The percentage germination of the pollen grains was 62% in distilled water. The pollen grains of *Amaryllis* are somewhat unusual in that they will germinate in water, without the addition of either glucose or sucrose. Whilst dilute solutions of both glucose and sucrose slightly increased the percentage germination, the primary effect of the sugar solutions was osmotic, growth of the grains being inhibited as the concentration of solute increased. Rapid growth of the grains occurred only in water or in a dilute sugar solution, hypotonic to the cell sap. Growth was markedly inhibited by a glucose solution containing 12 g glucose/100 cm³ and by a sucrose solution containing 21 g sucrose/100 cm³. At higher concentrations of glucose and sucrose, hypertonic to the cell sap, the pollen tubes either swelled and burst soon after emerging, or failed to germinate.

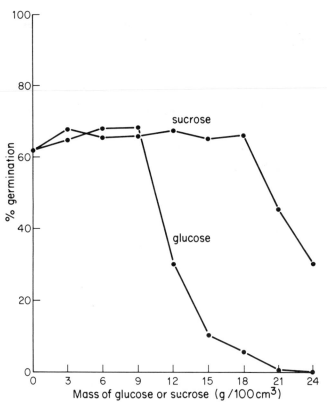

Fig. 77 *Percentage germination in pollen grains of a lily (Amaryllis hippeastrum) after immersion in solutions of glucose and sucrose*

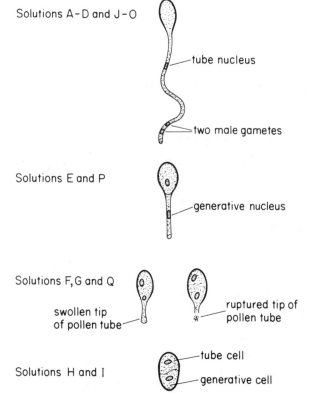

Fig. 78 *The appearance of pollen grains of Amaryllis hippeastrum in solutions of glucose and sucrose (see Table 68)*

126

3.12 A photometric method for measuring the growth rate of populations of yeast cells

2 See Table 69 and Fig. 79.

Table 69

Mass of dried yeast (g/100 cm³ water)	% absorbance of suspension
1.0	9
1.25	13
1.5	20
1.75	28
2.0	39
2.25	53
2.5	69
2.75	85
3.0	94

3 See Table 70 and Fig. 80.

Table 70

Period of incubation (h)	Dry mass of yeast (g)
0	1.0
6	1.1
12	1.23
18	1.26
24	1.4
30	1.6
36	1.76
42	1.9
48	2.1

3.13 An investigation into rates of fermentation by yeast

4 See Table 71 and Fig. 81 on page 128.

Table 71

Sugar	Distance travelled by meniscus (cm)					
	15	30	45	60	75	90 min
Glucose	0	1	10	17	48	96
Sucrose	0	0	1	5	22	57
Maltose	0	0	1	3	6	11
Lactose	0	0	0	0	0	0

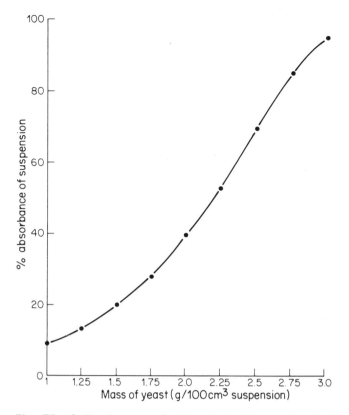

Fig. 79 *Calibration curve for determining the growth rate of yeast*

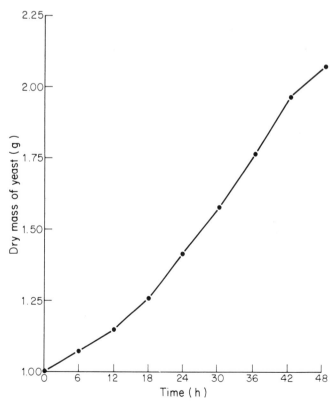

Fig. 80 *The growth rate of yeast in 0.5 M glucose solution at 35°C*

127

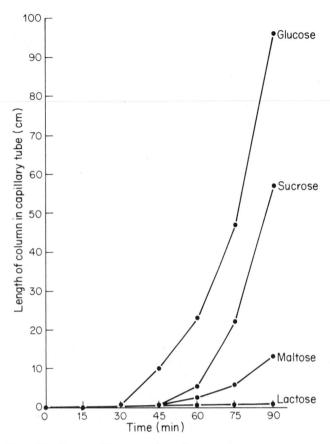

Fig. 81 *Rates of fermentation of sugars by yeast*

travel more than 30 cm in 15 minutes. The meniscus can, of course, be returned to the top of the capillary tube by raising the plunger of the syringe, but this is inconvenient and introduces a source of error. Alternatively, a wider tube could be used or the length of the capillary could be increased to 50 cm. Similarly, if weaker solutions of sugar had been used, or a less dense suspension of yeast, the rate of gas production would have been slower.

3.14 The effect of ethanol on the rate of anaerobic respiration of glucose by yeast

1 See Table 72. The graph (Fig. 82) is plotted from the reciprocal of time, $(1/t) \times 100$

Table 72

Temperature of incubation (°C)	Time for colour change in indicator (min)
20	40
30	20
40	11
50	no change

5 Yeast (*S. cerevisiae*) is capable of utilising glucose, sucrose and maltose as respiratory substrates, but this species of yeast cannot ferment lactose, as it lacks the enzyme necessary for the fermentation of this sugar.

Glucose was fermented most rapidly, possibly because this sugar can pass rapidly into the cell and enter directly into metabolic pathways. Sucrose, on the other hand, has to be hydrolysed by invertase into glucose and fructose before it can enter the cell.

$$\text{Sucrose} \xrightarrow{\text{Invertase}} \text{Glucose} + \text{Fructose}$$

Similarly, maltose must be hydrolysed by maltase before it can be fermented.

$$\text{Maltose} \xrightarrow{\text{Maltase}} \text{Glucose}$$

In addition, maltase is an inducible enzyme, secreted by yeast only when the substrate maltose is present in solution. Hence the relatively slow fermentation of maltose may be related to the fact that one of the enzymes involved in the fermentation of this sugar has to be induced.

As fermentation proceeds, bubbles of CO_2 gas are evolved, which rise to the top of the syringe barrel, where they collect and merge. Accumulation of carbon dioxide in the barrel causes displacement of an equivalent volume of liquid from the syringe barrel, which passes along the capillary, thereby providing a direct index of the rate of fermentation.

In the period 30–90 minutes after setting up the apparatus, evolution of carbon dioxide is so rapid that the meniscus may

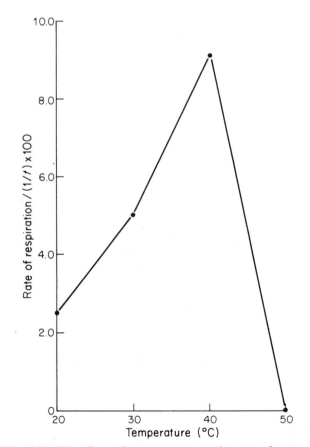

Fig. 82 *The effect of temperature on the rate of anaerobic respiration of glucose by yeast*

The Q_{10} of a reaction is defined as

$$\frac{\text{Rate of the reaction at } t + 10\,^{\circ}\text{C}}{\text{Rate of the reaction at } t\,^{\circ}\text{C}}$$

Hence, Q_{10} at 20–30 °C = 5.0/2.5 = 2.0

Q_{10} at 30–40° C = 9.1/5.0 = 1.82

Within the temperature range 20–40 °C the Q_{10} for respiration is approximately 2.0. At 50 °C, however, respiration was inhibited, which prevented any colour change in the indicator during the experiment.

2 See Table 73. The graph (Fig. 83) is plotted as a reciprocal of time, $(1/t) \times 100$.

Table 73

% ethanol by volume	Time for colour change in indicator (min)
0	20
1.5625	28
3.125	39
6.25	53
12.5	no change
25.0	no change

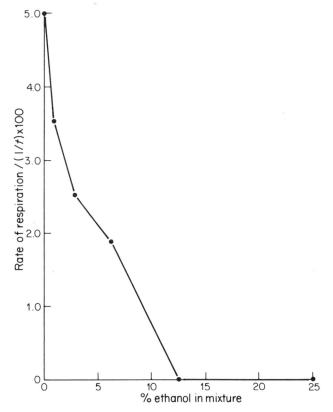

Fig. 83 *The effect of ethanol on the rate of anaerobic respiration of glucose by yeast*

Low concentrations of ethanol, within the range 1.5–6.25% v/v ethanol, depressed the rate of respiration in yeasts. Higher concentrations, within the range 12.5–25.0% v/v, inhibited respiration.

3.15 Arrangement and structure of vascular elements in petioles of celery

1 The red dye was observed to move upward through all vascular bundles of the petiole. Generally, the rate of ascent was rapid, varying from approximately 1 cm per minute in centrally positioned bundles to as much as 8 cm per minute in paired peripheral bundles, close to each margin of the petiole. Rates of ascent differed from bundle to bundle, reflecting the presence of xylem vessels of a different diameter in each bundle.

Walls of the xylem vessels were stained by the dye as it was carried upward in the transpiration stream. As time passed, the staining became more intense. Dye appeared in the leaves after about 10 minutes, whilst unabsorbed dye collected at the top of each vascular bundle, staining this region an intense red.

2 See Fig. 84.

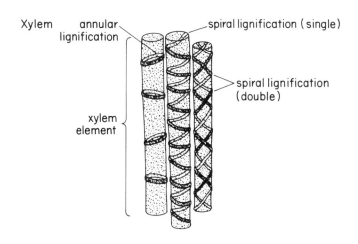

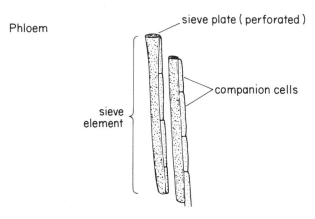

Fig. 84 *Conducting elements from the leaf petiole of celery*

4 See Fig. 85.

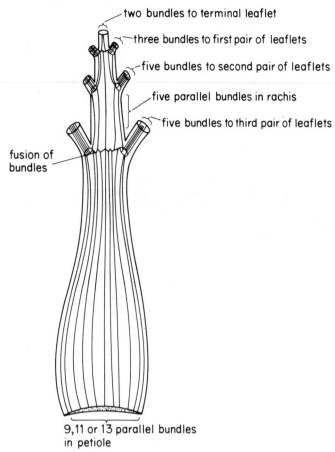

Fig. 85 *Arrangement of vascular bundles in the leaf petiole of celery*

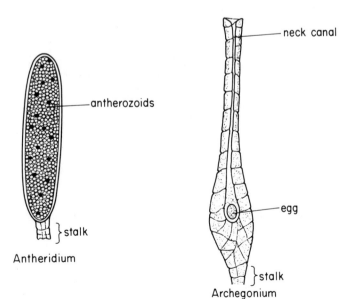

Fig. 86 *The structure of a mature antheridium and a mature archegonium of a moss*

3.16 Isolation of antheridia and archegonia from moss plants

1 See Fig. 86.

3.17 Variety of structure in epidermal trichomes

1, 2 and 3 See Fig. 87.

3.18 Morphology and cellular anatomy of four filamentous green algae

1 See Fig. 88.
2 See Fig. 89.
3 See Table 74.

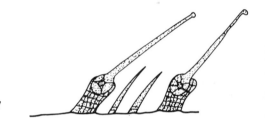

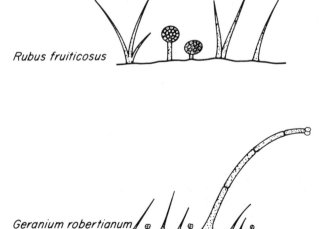

Fig. 87 *Variety of structure in epidermal trichomes*

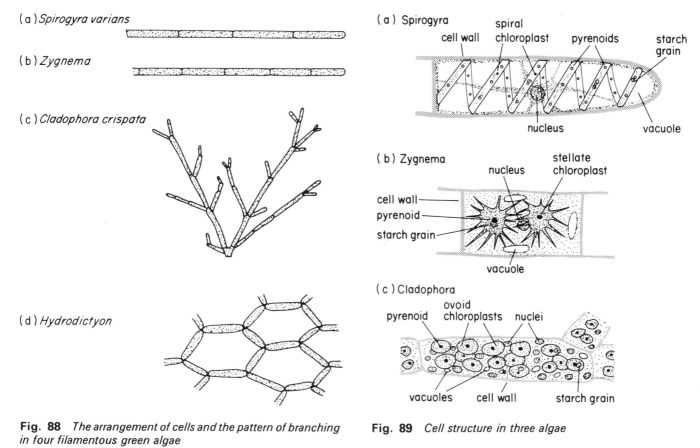

(a) *Spirogyra varians*

(b) *Zygnema*

(c) *Cladophora crispata*

(d) *Hydrodictyon*

(a) Spirogyra — cell wall, spiral chloroplast, pyrenoids, starch grain, nucleus, vacuole

(b) Zygnema — nucleus, stellate chloroplast, cell wall, pyrenoid, starch grain, vacuole

(c) Cladophora — pyrenoid, ovoid chloroplasts, nuclei, vacuoles, cell wall, starch grain

Fig. 88 *The arrangement of cells and the pattern of branching in four filamentous green algae*

Fig. 89 *Cell structure in three algae*

Table 74

Structural feature	*Spirogyra*	*Zygnema*	*Cladophora*	*Hydrodictyon*
Growth form	Unbranched filament	Unbranched filament	Dichotomously branched filament	A network of pentagons or hexagons, with each corner resulting from the union of three cells
Organisation of filament	Cellular, composed of cylindrical cells	Cellular, composed of cylindrical cells	A coenobium	A coenobium
No. nuclei per cell or unit	One	One	Many	One–many
No. and form of chloroplasts	One spiral chloroplast (Two or more in other species)	Two stellate chloroplasts	Many spherical-ovoid chloroplasts	Many band-like chloroplasts
Other structural features	Large vacuoles crossed by cytoplasmic strands	Small vacuoles	Units tend to swell at point of budding	Daughter nets may be present within older units

4.1 Active transport of chloride ions across the skin of a frog

5 See Table 75 and Fig. 90.

Table 75

Time (min)	Concentration of Cl⁻ ions in external solution	
	Outside-out skin	Inside-out skin
0	0	0
15	125	170
30	220	440
45	580	850

6 Chloride ions passed through the skin of the frog in both directions, but the rate of transport was more rapid from the outside to the inside than from the inside to the outside. This implies the existence of an active mechanism, probably related to the fact that the frog lives in fresh water, which is hypotonic to its body fluids, and can accumulate chloride ions, and retain them within its body, against a concentration gradient.

It should be noted that a 0.5 M solution of sodium chloride, which is toxic to living cells of the skin, probably kills the skin within a period of 30–40 minutes. In dead skin, of course, the rate of chloride ion transport is the same in both directions.

4.2 An investigation of the action of enzymes in the alimentary canal of a freshly killed mammal

1 See Table 76.

Table 76

Organ	pH
Oesophagus	6.0
Stomach	3.5
Duodenum	7.5
Mid-small intestine	6.8
Ileum	6.5
Colon	6.0

6 See Table 77.

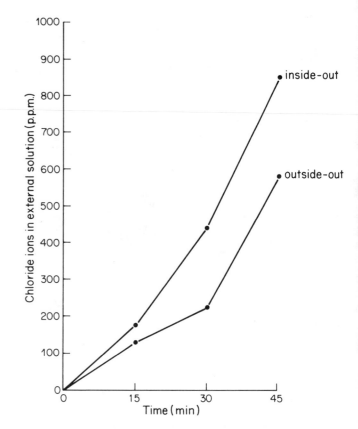

Fig. 90 *Transport of chloride ions across the skin of a frog*

Table 77

Organ	Enzymes					
	Amylase	Protease	Lipase	Invertase	Maltase	Lactase
Oesophagus	✓ (118 mm²)	—	—	—	—	—
Stomach	✓ (102 mm²)	✓ (124 mm²)	✓ (114 mm²)	—	—	—
Duodenum	✓ (168 mm²)	✓ (148 mm²)	✓ (136 mm²)	✓	✓	✓
Mid-small intestine	✓ (152 mm²)	✓ (136 mm²)	✓ (128 mm²)	✓	✓	✓
Ileum	✓ (146 mm²)	✓ (130 mm²)	✓ (124 mm²)	✓	✓	✓
Colon	✓ (106 mm²)	—	—			

132

Dietary starch is digested by two enzymes, notably amylase and maltase. Amylase is secreted into the alimentary canal by the salivary glands and pancreas. Hence amylase, originating from either of these sources, was found to be present in the oesophagus and throughout the small intestine. Additionally, amylases may be produced by sub-mucosal tissues of the stomach and colon. These amylases are released when excised portions of organs are transferred to the starch-agar plate. Maltase, invertase and lactase are secreted in the succus entericus and were found to be active throughout the entire length of the small intestine.

Dietary proteins are digested by enzymes secreted from three regions of the alimentary canal, notably the stomach (pepsin), pancreas (trypsin and related enzymes) and small intestine (erepsin and related enzymes). Proteolytic enzymes were found to be present in the stomach and throughout the small intestine of the rat, but were absent from the oesophagus and colon.

Dietary fats are digested by lipases secreted by the stomach, pancreas and small intestine. As in the case of proteins, lipase activity was found to be confined to the stomach and small intestine.

4.3 An investigation of human sensory receptors

1 When dry cotton wool was stroked over (i) the palm of the hand and (ii) the forearm, a tickling sensation was experienced in both regions. This sensation was, however, more noticeable in the hair-covered forearm than in the hairless palm, chiefly because touch receptors at the base of hairs are stimulated by light contact.

2 (i) The primary sensation was one of intense heat, caused by stimulation of heat receptors, which are particularly numerous in the region of the palm. Later a tingling sensation was experienced.
  (ii) The sensation was one of heat and of pain, caused by stimulation of heat and pain receptors in the back of the hand. The sensation of heat is less intense than in the palm, reflecting the presence of relatively fewer heat receptors in this area of the skin.
  (iii) Rubbing the forearm produced a sensation of mild heat. Heat receptors are present but are less numerous here than in the palm or back of the head.

3 (i) Tip of thumb.
  (ii) Lips.
  (iii) Forehead.
Ability to detect the two pinheads depends on the density of touch receptors in different areas of the skin. Only in those areas listed are the touch receptors sufficiently numerous for most subjects to detect two distinct points of contact.

4 Forearm $\longrightarrow$ palm $\longrightarrow$ back of hand $\longrightarrow$ face.

5 (i) Sodium chloride is tasted in region three.
  (ii) Sucrose is tasted in region three.
  (iii) Lemon juice is tasted in region two.
  (iv) Quinine is tasted in region one.
None of these compounds is tasted in region four. Water, however, is tasted in region three; recent research has indicated that in addition to salt, sweet, sour and bitter, the human tongue is also capable of tasting water.

6 Focal lengths vary from individual to individual and sometimes between the left and right eyes of an individual.

Generally, for this test, focal lengths lie within the range 20–50 cm.
  (i) At a particular distance from the eye, one of the circles disappeared from view because light rays from this circle fell on the blind spot of the eye, which has no light-sensitive cells.
  (ii) The distance at which the circle disappeared from view measures the focal length of the eye.
  (iii) Any increase in the focal length of the eye is most probably related to farsightedness (hypermetropia), which results in a blurred image, unless corrected by a convex lens.

7 The eye can be accommodated either for near vision or for distant vision, but not for both. If the pen is brought into sharp focus, then the eye is accommodated for near vision, and the corner of the room is out of focus. Conversely, when the corner of the room is in sharp focus, with the eye accommodated for distant vision, the outline of the pen is indistinct.

8 The left ear of each subject would be blocked with cotton wool. Subjects would be asked to place their right ear at the edge of the bench. The clock would then be moved along the bench surface towards the subject, until a ticking sound could be heard. A mark would be made at the point where the sound first became audible, and the distance between this mark and the edge of the bench used as an index of the relative efficiency of hearing. The procedure would then be repeated to assess the relative efficiency of hearing in the left ears of subjects.

4.4 Rates of glucose absorption from different regions of the alimentary canal in a freshly-killed mammal

4 See Table 78 and Fig. 91.

Table 78

Time (min)	Concentration of glucose in external solution (g/100 cm³)			
	Stomach	Duodenum	Ileum	Large intestine
0	0	0	0	0
15	0	0.25	0	0
30	0	0.5	0.25	0.25
45	0	0.5	0.25	0.25
60	0	0.75	0.5	0.25
75	0	0.75	0.5	0.25
90	0.25	0.75	0.5	0.5

5 As a result of using a freshly-killed mammal and of transferring excised regions of the alimentary canal to isotonic saline, it is likely that many of the cells in the four regions of the gut retained their normal functions for most, if not for all, of the experiment.
Stomach There was little, if any absorption of glucose from the lumen of the stomach. This may have resulted from a physical effect, dependent on the thickness and muscular nature of the stomach wall, or it may have resulted from a physiological effect, in which the lining cells inhibited the absorption of glucose from the lumen.

133

Duodenum Glucose was rapidly absorbed from this region. In the living animal this is known to be the principal region from which glucose is absorbed, probably as a result of an active process.

Ileum Some glucose was absorbed from this region, but the rate of absorption was slower than in the duodenum. There is no obvious explanation for this, although active transport of glucose may not occur from this region of the small intestine.

Large intestine Some glucose was absorbed from the rectal region of the large intestine, in which the wall is relatively thin.

6 (i) No attempt was made in this experiment to measure rates at which glucose was absorbed through comparable surface areas of the wall of the alimentary canal. Whilst some of the excised regions of the alimentary canal were of roughly comparable length, none were of comparable surface area.

(ii) It may be that glucose is absorbed from some regions of the alimentary canal as a result of a rapid, active process. In order to determine if this is a correct interpretation, a control must be set up, consisting either of a comparable region of the gut, killed by immersion in hot water, or a non-living semi-permeable membrane, such as Visking tubing.

4.5 To determine the amount of urea in urine

4 See Fig. 92. Amounts of urea in urine, estimated for a class of 20 students, gave results of between 0.5 and 2.5 g urea/100 cm³ urine.

5 The calibration curve for urea is probably a straight line, providing the substrate (urea) is present in excess. Whilst this method appears to produce very accurate results for concentrations of urea within the range 2–8 g/100 cm³, it is less accurate when used for the estimation of more dilute solutions of urea.

Amounts of urea in urine generally lie within the range 0.5–2.5 g/100 cm³, but the figure is variable, depending on a number of factors. Levels of urea may rise following consumption of a protein-rich meal, or fall after consumption of a large volume of water.

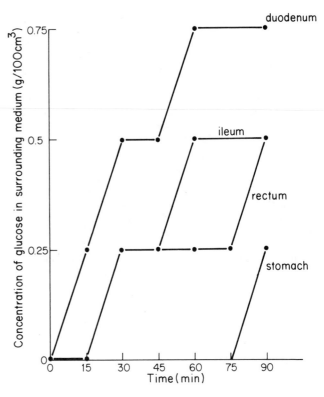

Fig. 91 *Rates of glucose absorption through excised regions of the alimentary canal of a freshly killed rat*

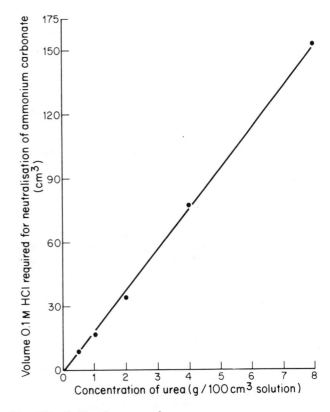

Fig. 92 *Calibration curve for urea*

4.6 Self assessment of cardiac and renal efficiency

1 See Table 79 and Fig. 93.

Table 79

Time (min)		Rate of heartbeat (beats/min)
−2 −1	period of preparation	70 72
0 1 2	period of exercise	not measured not measured not measured
3		126
4		97
5		88
6		76
7		73
8		71
9		71
10		71

The increase in the rate of the heartbeat during exercise provides an index of the efficiency of the cardiovascular system.

The slight rise in the rate of heartbeat immediately before the commencement of the exercise is the result of the secretion of adrenaline from the adrenal glands and nor-adrenaline from the sympathetic nervous system.

2 (i) Providing the amount of work undertaken is standardised, an increase in the heart rate implies decreased efficiency of the cardiovascular system in supplying blood to the tissues. The cardiovascular system is under greater stress and the heart responds by beating at a faster rate.

   (ii) Again, providing the amount of work undertaken is standardised, a decrease in the heart rate implies improved efficiency of the cardiovascular system. If the rate of heartbeat was found to be slower than on some previous occasion, the efficiency of the heart as a pump must have increased, or peripheral resistance to blood flow in the capillaries must have decreased.

3 In the method used for measuring pulse rate, it is not possible to measure the rate of heartbeat during the period of exercise, nor to plot the rate of increase to the maximal level. These problems could be overcome by connecting the subject to an electrocardiograph.

4 Total volume of urine = 195 cm³
Sodium chloride content of the urine = 0.58 g/100 cm³.

5 750 cm³ water was consumed. The results are given in Table 80 and Fig. 94 on page 136.

6 Following consumption of a large volume of water, there was a subsequent increase in the volume of the urine, which was produced according to a characteristic pattern in all subjects tested. Generally, more than one-half of the total volume of the water consumed was excreted within the first 90 minutes, while the remainder (minus 200–300 cm³ lost in sweat or in the breath), was excreted over a period of 2–3 hours. The greater the efficiency of the kidney, the faster will it remove excess water from the body.

After drinking a large volume of water, and in response to the resulting dilution of the blood, output of antidiuretic hormone from the posterior pituitary gland is reduced. The kidneys respond by producing a copious urine with a low specific

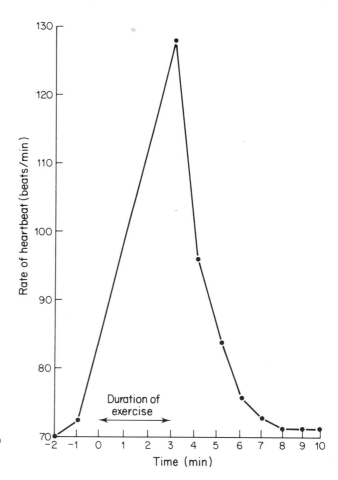

Fig. 93 *The effect of exercise on the rate of heartbeat*

Table 80

Time after drinking (min)	Volume urine passed (cm³)	Sodium chloride content of urine (g/100 cm³)	Total sodium chloride excreted (g)
0	0	0	0
30	57	0.50	0.28
60	126	0.12	0.151
90	186	0.076	0.141
120	115	0.132	0.152
150	61	0.25	0.152
180	0	0	0

gravity. Levels of sodium chloride in the urine, 0.58 g/100 cm³ before drinking, fell to 0.12 g/100 cm³ after 60 minutes and recovered to 0.25 g/100 cm³ after 3 hours. It seems probable that as a result of drinking in excess of requirement, more sodium chloride was eliminated from the body than if drinking had been maintained at normal levels.

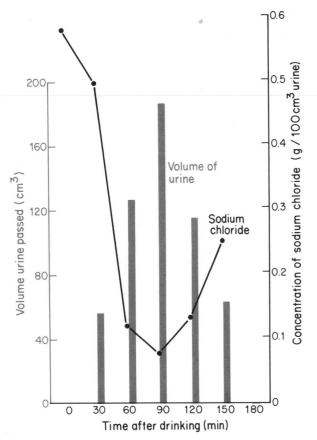

Fig. 94 *The effect of consuming 750 cm³ water on urine output and sodium chloride content of urine*

4.7 The effects of stimulants and depressants on the rate of heart beat in water fleas

1 and 2 See Table 81 and Fig. 95.

Table 81

Time (min)	Rate of heart beat (beats/min)				
	Water	Ethanol	Adrenaline	Chlor-promazine	Salicylic acid
0	184	184	184	184	184
2	188	120	220	86	174
4	189	120	232	60	160
6	190	140	240	48	154
8	192	145	244	24	151
10	196	150	232	18	138

3 *Distilled water* When water fleas were immersed in distilled water there was a slow, steady rise in the rate of heart beat, possibly related to oxygen deficiency in the water.
Ethanol Ethanol is a depressant, which caused a rapid initial fall in the heart rate. In many animals, however, the heart rate then recovered, implying a fairly low toxicity.

Adrenaline Adrenaline is a stimulant, which caused an almost immediate rise in the rate of heart beat of around 240 beats per minute. This high rate, however, was not maintained, possibly as a result of the toxicity of adrenaline.
Chlorpromazine Chlorpromazine is a powerful depressant and toxin, which markedly slowed the heart rate, and in many cases caused death by cardiac arrest.
Salicylic acid Salicylic acid is also a depressant and toxin, which exerted an effect similar to that of chlorpromazine, although the toxicity was less marked.

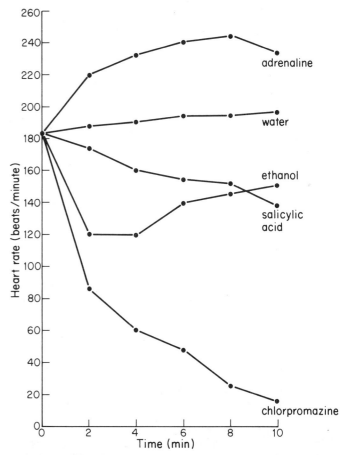

Fig. 95 *The effect of stimulants and depressants on the rate of heart-beat in water fleas*

4.8 Factors affecting the vertical distribution of water fleas in water

1 See Table 82.

Table 82

Region	Numbers of *Daphnia*			
	0	5	10	15 min
Upper	21	24	25	27
Middle	23	14	12	9
Lower	28	34	35	36

The distribution of animals, which was roughly even immediately after filling the tubing, became uneven, with aggregations developing in both the upper and lower regions of the water column, close to each air-water boundary. Relatively few animals took up a position in the middle region of the water column.

The uneven distribution of animals may be related wholly or partly to:

(i) an oxygen gradient in the water,
(ii) the presence of a food source both at the surface and at the bottom of the water column, or
(iii) an innate response to gravity—an unlikely cause in view of the observed distribution.

2 The effect of illuminating the lower region and of shading the upper region, is shown in Table 83.

Table 83

Region	Numbers of *Daphnia*			
	0	5	10	15 min
Upper	27	28	28	30
Middle	9	10	12	12
Lower	36	34	32	30

Light intensity does not therefore appear to be a major factor in determining the distribution. Generally, the water fleas were tolerant of high light intensity, but a small number migrated from the illuminated region and did not return. The majority of these migrants moved through the middle region to take up a position in the upper region.

3 See Table 84.

Table 84

Region	Number of *Daphnia*			
	0	5	10	15 min
Upper	29	28	26	24
Middle	10	8	7	6
Lower	33	36	39	42

Yeast, which is a food source for the *Daphnia*, attracted a number of animals from the upper and middle regions into the lower region, where they remained to feed.

4 See Table 85.

Table 85

Region	Numbers of *Daphnia*			
	0	5	10	15 min
Upper	27	29	32	33
Middle	10	12	13	12
Lower	35	31	27	27

One effect of olive oil was to prevent absorption of oxygen at the lower end of the water column. As a result of this, there appeared to be a slow migration of animals from this region towards the upper region, where more oxygen was available.

5 Although there appears to be a definite aggregation of animals in the upper and lower regions of the water column, there is a need to apply tests of significance in order to evaluate the results.

If the animals do generally aggregate in the upper and lower regions of the water column, then neither light intensity nor gravity appear to be environmental factors which exert a primary influence on the distribution. Further experiments need to be carried out: (i) using more intense illumination, applied for longer periods of time to the upper, middle and lower regions, and (ii) with the tube positioned horizontally, in order to determine if gravity has any influence on the distribution.

Generally, there is a need to continue with the counts over a longer period of time. The results that were obtained, however, appear to imply that the distribution was determined primarily by the nutritional and respiratory needs of the animals, which migrated to the ends of the water column where both food and oxygen were more abundant. A microscopic examination of water samples from both ends of the column would show if food was present. If the experiment was repeated using distilled water, and a similar result was obtained, this would point to the oxygen gradient as the major environmental factor responsible for the distribution.

4.9 An investigation into behaviour of maggots of the blow-fly

1 See Table 86.

Table 86

Substratum	Time required for maggot to travel 15 cm (s)			No. gut pulsations		
	1st run	2nd run	Mean	1st run	2nd run	Mean
Writing paper	48	44	46	44	42	43
Plastic sheeting	56	52	54	58	50	54
Sand paper	62	54	58	60	64	62

2 If a maggot travels 15 cm in 46 seconds, then its rate of locomotion on writing paper is

$$\frac{15}{46} = 0.326 \, \text{cm s}^{-1}$$

and on plastic sheeting

$$\frac{15}{54} = 0.278 \, \text{cm s}^{-1}$$

and on sand paper

$$\frac{15}{58} = 0.259 \, \text{cm s}^{-1}$$

3 Apparent pulsating movements of the gut result from alternate contraction and relaxation of the muscles used in locomotion. On writing paper, which the maggot can grip and which does not interfere with locomotion, the rate of the pulsations is directly related to the rate of locomotion. On the other hand, the rate of pulsations is not related to the rate of locomotion on either plastic, which the maggot cannot grip, or sand paper, which presents an uneven, rough surface.

4 The maggot moved to one side of the corridor, then maintained lateral contact with one of the blocks of wood. The response is called positive thrigmotaxis. This adaptive behaviour is designed to reduce water loss, which would be greater if the entire skin surface were exposed to the drying effect of the air.

5 See Fig. 96.

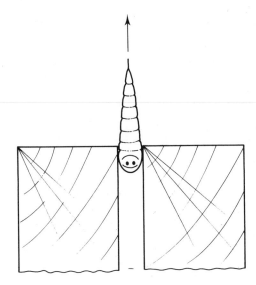

Fig. 97 *Response of a maggot to bilateral contact. On emerging from between the wooden blocks, the maggot continued in a straight line*

Table 87

Time (min)	No. maggots per compartment					
	Not illuminated		50 cm from lamp		10 cm from lamp	
	Light	Dark	Light	Dark	Light	Dark
0	30	0	30	0	30	0
1	25	5	12	18	6	24
2	14	16	6	24	1	29
3	8	22	2	28	0	30
4	9	21	1	29	0	30
5	9	21	1	29	0	30
6	6	24	2	28	2	28
7	7	23	3	27	3	27
8	6	24	2	28	3	27

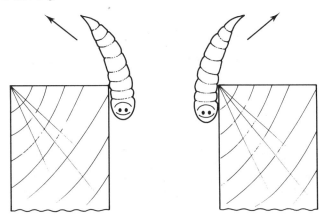

Fig. 96 *Response of a maggot to unilateral contact. On reaching the end of the wooden block, the maggot turned in the direction of the stimulated side*

6 See Fig. 97.

7 See Table 87 and Fig. 98. Generally, as the light intensity was increased, so more maggots moved rapidly from the illuminated compartment into the darkened compartment, but at lower light intensities some maggots re-entered the illuminated compartment from the darkened compartment.

8 (i) Materials
- Blow-fly maggots
- 2 wooden blocks, as supplied
- A stop watch
- A bench lamp
- Metre rule

(ii) Method
(a) Place the two wooden blocks on the bench surface, parallel to one another and approximately 1.5–2.0 cm apart.

(b) Using the metre rule, mark positions on the bench surface at 10, 20, 30, 40 and 50 cm from one end of the wooden blocks.

(c) Position a maggot at one end of the corridor between the blocks facing down the corridor.

(d) Place the lamp at 10 cm from the blocks, behind the maggot. Switch on the lamp and record the time taken for the maggot to complete the run. Obtain a second set of results. Convert readings to centimetres per second.

(e) Obtain results with the lamp positioned at 20, 30, 40 and 50 cm from the wooden blocks.

(f) The intensity of illumination should be expressed as a reciprocal of the distance of the lamp from the wooden blocks. A convenient formula is $(1/d^2) \times 1000$, where d = distance of the lamp (cm) from the blocks.

(g) Plot a graph of rates of locomotion against intensity of illumination.

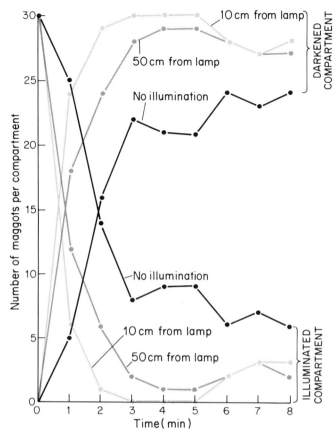

Fig. 98 *Movement of maggots between illuminated and darkened compartments of the petri dish*

4.10 The effect of light intensity and light quality on the rate of locomotion in maggots of the blow-fly

1 See Table 88.

Table 88

Quality of light	Trial no.	\multicolumn{5}{c}{Time taken to move around one-half of dish(s)}
		1	2	3	4	Mean
White light		32	33	32	33	32.5
Darkness		45	45	51	48	47.5

The rate of locomotion was 46.15 per cent slower in darkness than in white light. This result suggests that the maggot is negatively phototactic, its rate of locomotion increasing as light intensity is increased.

2 and 3 See Table 89.

Percentage reduction in the rate of locomotion caused by each filter:

Yellow	14.4
Red	30.6
Blue	25.4
Pea green	16.9
Dark green	35.4

Table 89

Quality of light	Trial No.	\multicolumn{5}{c}{Time taken to move around one-half of dish(s)}
		1	2	3	4	Mean
Yellow		37	39	36	39	37.75
White		33	34	32	33	33.0
Primary red		41	40	40	41	40.5
White		30	31	32	31	31.0
Light blue		43	42	37	41	40.75
White		32	33	33	32	32.5
Pea green		42	34	35	34	36.25
White		30	31	32	31	31.0
Dark green		41	44	48	43	44.0
White		32	33	32	33	32.5

4 See Table 90 and Figs 99 and 100.

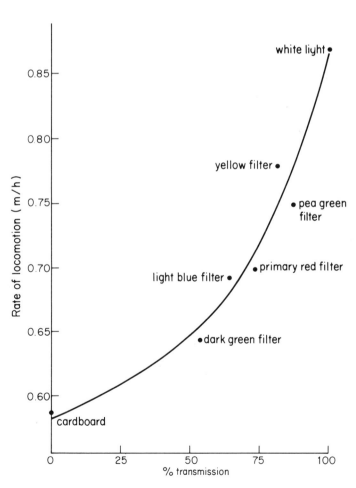

Fig. 99 *The relationship between light intensity and the rate of locomotion in maggots of the blow-fly*

Table 90

Quality of light	Rate of locomotion (cm s⁻¹)
White	0.858
Yellow	0.749
Red	0.700
Blue	0.694
Pea green	0.180
Dark green	0.642
Darkness	0.595

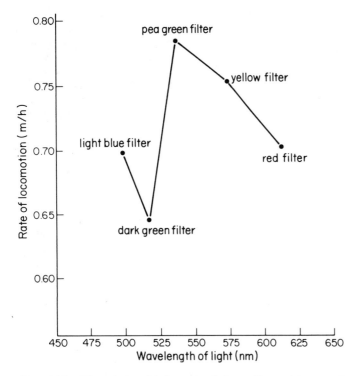

Fig. 100 *The relationship between light quality and the rate of locomotion in maggots of the blow-fly*

5 In general, the rate of locomotion in maggots appeared to be directly related to light intensity, increasing as light intensity increased. Conversely, there were no general indications that the rate of locomotion was affected by the wavelength of the light falling on the animals. Even so, it may be that a weak response was shown to light at the blue and red ends of the spectrum, blue light decreasing the rate and red light accelerating it.

Further investigations could be carried out by:
(i) using filters of different thicknesses, each with the same peak of transmittance, and
(ii) using a wider range of filters, each with different peaks of transmittance, but with the same percentage absorbance of white light.

4.11 Aspects of behaviour in the periwinkle

1 The experiment is designed to determine if the periwinkles show a negative geotactic movement in (i) air and (ii) water, and if the response is modified in any way by light or darkness.

Periwinkles surrounded by air remained in position at the bottom of the cylinder. When submerged beneath water, however, some animals made a slow ascent (that is, a negative geotactic response) which was particularly marked in those animals previously exposed to air. Generally, more animals ascended in darkness than in light, but even in darkness some animals remained at the bottom of the cylinder.

Periwinkles live on rock surfaces, feeding on algae. At low tide, during the day, they often remain motionless, adhering to the rock surface, camouflaged by the vegetation. The strong negative geotaxis shown by the majority of individuals in water may cause them to climb to the high points of rocks, where they can breath air as the tide ebbs. At night the animals tend to be more active, moving freely over the rock surfaces.

2 Periwinkles previously kept in darkness exhibited a strong positive phototactic movement, which was most evident after 10 minutes. The response, however, weakened with time. After 20–30 minutes some animals showed a distinct reversal of the response, moving directly away from the light source.

The response may play a role in determining the distribution of periwinkles on rock surfaces, causing the animals to occupy illuminated areas as the tide ebbs. These illuminated areas are generally the highest points of rock surfaces where edible algae are to be found.

3 See Table 91 and Fig. 101.

Table 91

% sea water in mixture	No. emergent periwinkles
0	0
10	0
20	0
30	0
40	0
50	4
60	6
70	6
80	7
90	7
100	8

At concentrations below 50% sea water, the periwinkles remained enclosed within their shells. In 50% sea water some individuals emerged from their shells, and the number of emergent animals increased as concentrations of sea water were increased. Even so, in undiluted sea water some animals remained within their shells.

4 See Table 92 and Fig. 102. The graph is plotted as a median value of the force required to detach each periwinkle from the substratum. Generally, there is a positive correlation between adhesive force and body size, but the correlation is not absolute. Some animals show disproportionate strength or weakness, an effect that may be determined by environmental factors, such as the position on the shoreline occupied by each animal, and the force of the waves with which it has had to contend.

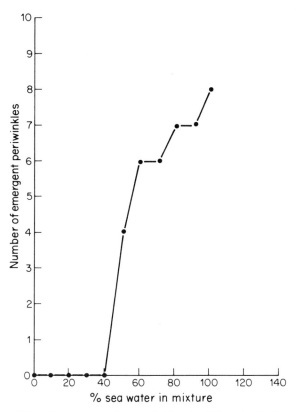

Fig. 101 *Emergence of periwinkles in different dilutions of sea water*

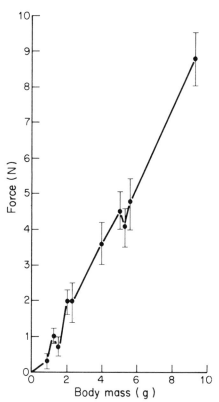

Fig. 102 *The relationship between the mass of individual periwinkles and the force required to pull them from the substratum*

Table 92

Body mass of periwinkle (g)	Force required to detach animal(N)
0.8	0.1–0.5
1.2	0.8–1.2
1.5	0.5–1.0
2.0	1.6–2.3
2.3	1.4–2.5
4.0	3.0–4.2
5.0	4.0–5.1
5.3	3.5–4.6
5.6	4.0–5.4
9.2	8.0–9.5

5 It would be necessary to collect at least 50 periwinkles from (i) the high water mark and (ii) the low water mark. Individual periwinkles from each batch would be weighed and their masses recorded. After killing the animals, and removing their shells, shells would be weighed and their masses recorded.

The index: $$\frac{\text{mass of shell (g)}}{\text{mass of whole animal (g)}}$$

would be used to test the hypothesis. Results would be subjected to statistical analysis to determine if differences between the two batches were significant.

4.12 Locomotion and behaviour in larvae of the mosquito

1 Undisturbed larvae ascended by making undulating movements of the abdomen, with the head pointed downwards. In each movement the abdomen was bent, first to one side, then to the other, while the tip of the abdomen was rotated. After taking air at the surface, undisturbed larvae often descended head first, either with the body held straight, or coiled into a semi-circle (Fig. 103). As each larva is slightly more dense than the surrounding water, it slowly sinks to the bottom of the vessel under the influence of gravity.

The chief effect of tapping the side of the vessel was to cause the larvae to dive. In this movement the head was held uppermost and the abdomen moved from side to side in a similar manner to that used for ascent through water.

2 Small larvae made more swimming movements per minute than large ones, and there appeared to be a direct relationship between body size and the number of movements made (Table 93).

Table 93

Length of larva (mm)	No. movements/min
2	108
5	89
8	68

Although the larger larvae made fewer movements per minute, they nevertheless travelled through the water at a faster rate (Table 94).

141

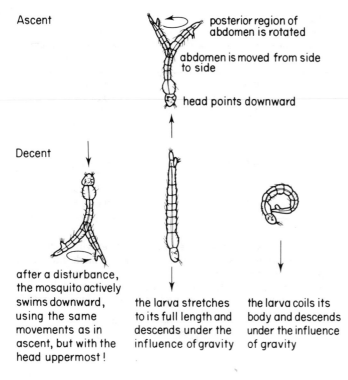

Ascent

posterior region of abdomen is rotated

abdomen is moved from side to side

head points downward

Decent

after a disturbance, the mosquito actively swims downward, using the same movements as in ascent, but with the head uppermost!

the larva stretches to its full length and descends under the influence of gravity

the larva coils its body and descends under the influence of gravity

Fig. 103 *Methods of locomotion in larvae of the mosquito*

Table 94

Length of larva (mm)	Distance travelled (cm/min)
2	13
5	33
8	65

3 See Fig. 104. If they were not disturbed, the larvae spent most of the time at the surface of the water, breathing air. Occasionally, and generally not more than once or twice in each period of 20 minutes, each larva left the surface and moved downward through the water, remaining submerged for not more than 1–3 minutes.

   If the temperature of the water had been raised by 10 °C, levels of oxygen in the water would have been reduced. The larvae would probably have spent more time at the surface of the water and less time submerged.

4 See Fig. 105. The mosquito larvae dived in response to the presence of cardboard and to vibration of the vessel. Whilst the initial reaction was marked, with almost all of the larvae diving to below the 50 cm^3 mark, the response became less marked with each repetition. This type of response is known as habituation. After the mosquito larvae have become fully habituated to the stimulus, they no longer dive.

   After a resting period of 20 minutes, however, the intensity of the reaction would have recovered to its original level.

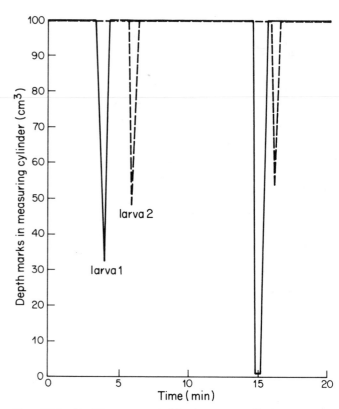

Fig. 104 *Positions occupied by two mosquito larvae over a period of 20 minutes*

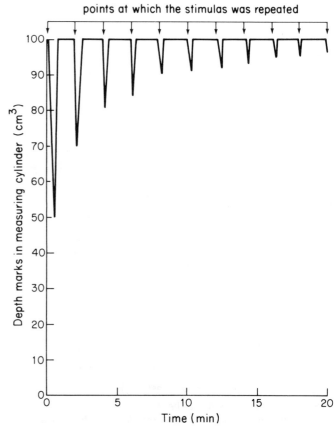

Fig. 105 *Habituation in larvae of the mosquito*

4.13 Aspects of osmoregulation and behaviour in the shore crab

1 See Table 95.

Table 95

Time (min)	Titrator reading	Mass sodium chloride (g/100cm³)	Chloride in concen- tration (p.p.m.)
0	0	0	0
10	3.4	0.029	171
20	4.6	0.047	279
30	5.2	0.058	349
40	5.6	0.067	404
50	5.8	0.073	435
60	5.9	0.075	451

2 See Fig. 106. The graph is plotted from the loss of chloride ions recorded in each 10 minute period against time. By extrapolation of the curve to the x-axis, it is possible to predict that an ionic equilibrium, at which no further losses of sodium chloride will occur, would be established after 70–80 minutes.

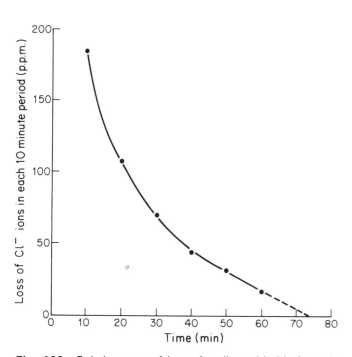

Fig. 106 *Relative rates of loss of sodium chloride from the gills of a shore crab following immersion in distilled water*

3 (i) The crab should have been washed in distilled water, to remove salt from the body surface, before it was placed into the jar.
  (ii) At intervals of 10 minutes throughout the experiment, 2.0 cm³ of water are withdrawn from the original volume of 100 cm³. Therefore, apart from the first reading, the concentrations of sodium chloride recorded are not measurements of the mass of solute contained in 100 cm³ solution.

4 The crab burrowed into the sand, digging with legs 2 to 5. It stopped digging when the carapace was level with, or just below, the surface of the sand. In this position the crab remained motionless, with the stalked eyes protruding above the sand. This behaviour enables the crab to avoid predators in those regions of the shore that are sandy.

5 The crab showed a positive response to the strip of black paper, moving towards the paper and generally resting on it, regardless of whether it was placed to the left or to the right of the animal.

This behaviour probably helps the crab to avoid predators, by causing it to move towards dark rock crevices, in which it can hide, on rocky regions of the sea shore.

6 In response to tapping the posterior part of the carapace, crabs moved to the left, the right, or forward, but rarely, if ever, backward. Movement was effected by raising the body on legs 2 to 5, the first pair of clawed legs playing no part in movement.

Individual crabs tended to make lateral movements that were predominantly, but not exclusively, in one direction, both on land and in water. Results presented in Table 96 were typical of individual animals.

Table 96 L = moved to left, R = moved to right, F = moved forward.

Trial	Land	Water
1	L	L
2	L	L
3	L	L
4	L	L
5	L	L
6	L	L
7	F	L
8	R	L
9	R	L
10	R	L

(It should be noted that in the trials on land a change in the direction of lateral movement was preceded by a forward movement. This result was observed in some, but not in all, crabs tested.)

4.14 Removal of organ systems from a rat

3 (i) See Fig. 107.
   (ii) See Fig. 108.
   (iii) See Fig. 109.
   (iv) See Figs 110 and 111.

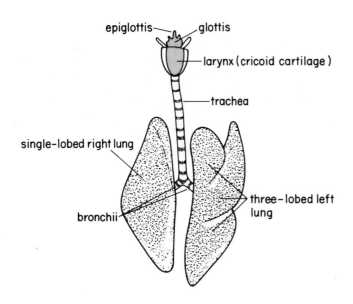

Fig. 107 *The respiratory system of a rat*

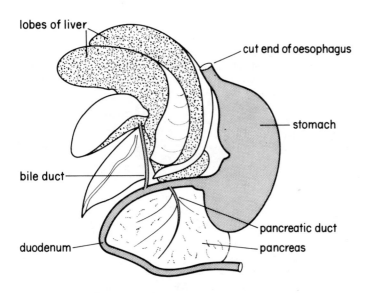

Fig. 108 *The stomach, duodenum and associated glands of a rat*

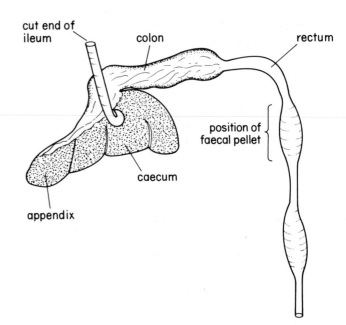

Fig. 109 *The large intestine of a rat*

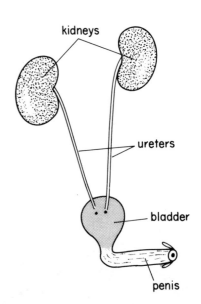

Fig. 110 *The excretory system of a male rat*

144

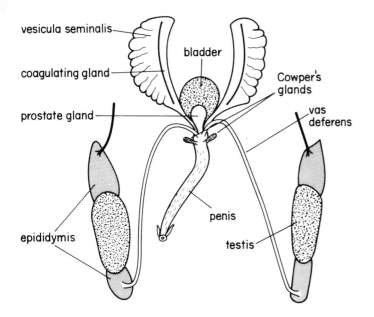

vesicula seminalis

coagulating gland

prostate gland

epididymis

bladder

Cowper's glands

vas deferens

penis

testis

Fig. 111 *The reproductive system of a male rat*

Further Reading

Note: Photocopies of articles from learned journals, scientific texts etc., intended for use in private research, are available from the Photocopy Service, Science Museum Library, London SW7 5NH. Requests should be made on Form 197 ScM and accompanied by the appropriate remittance.

Andrew, B. L. (1969) *Experimental Physiology*. E. & S. Livingstone, London.

Andrews, I. M. (1975) A method for the characterisation of enzymes. *School Science Review*, 56 (196), 534–536.

Andrews, R. J., Owen, R., Swann, D. and Waterman, K. H. (1979) Seashore ecology in the sixth form biology course. *School Science Review*, 60 (212), 503–511.

Brown, C. R. (1981) Some uses in school biology for the prawn. *School Science Review* 62 (221), 704–708.

Coote, N., Kirsop, B. H. and Buckee, G. K. (1973) The concentration and significance of pyruvate in beer. *J. Inst. Brewing*, 79, 298–304.

Cunningham, A. W. (1979) A simple experiment to find the rate of water uptake in a shoot. *School Science Review*, 61 (215), 274–75.

Dam Kofoed, A. and Kjellerup, V. (1970) Movements of fertiliser nitrogen in soil. *Tidesskrift fur Planteave*, 73, 659–686.

Dijkshoorn, W. and Ismunadji, M. (1972) Nitrogen nutrition of rice plants measured by growth and nutrient content in pot experiments. 2. Uptake of ammonium and nitrate from a waterlogged soil. *Neth. J. Agric. Sci.*, 20, 44–57.

Dixon, N. E., Gazzola, C., Blakeley, R. L. and Zerner, B. (1975) Jack bean urease, a metalloenzyme. *J. Am. Chem. Soc.* 97, 4131–4132.

Freeland, P. W. (1971) A photometric method for determining the growth rate of yeast. *Journal of Biological Education*, 5, 141–143.

Freeland, P. W. (1973) Some practical aspects of sugar fermentation by baker's yeast (*S. cerevisiae*). *Journal of Biological Education*, 7, 14–22.

Freeland, P. W. (1973) Some applications of glucose-sensitive reagent strips in biology teaching. *School Science Review*, 55 (190), 91–96.

Freeland, P. W. (1974) Characterisation of digestive enzymes in freshly killed animals. *Journal of Biological Education*, 8, 38–45.

Freeland, P. W. (1974) Catalase induction in baker's yeast. *Journal of Biological Education*, 8, 195–200.

Freeland, P. W. (1974) Some applications of agar-gel diffusion techniques. *School Science Review*, 56 (195), 274–287.

Freeland, P. W. (1975) Some applications of reagent strips in soil experiments. *School Science Review*, 56 (197), 738–741.

Frias, F. (1972) *Practical Biochemistry, an Introductory Course*. Butterworths, London.

Gale, J. and Hagan, R. M. (1966) Plant antitranspirants. *Ann. Rev. Plant Physiol.* 17, 269–282.

Goulding, K. H. and Merrett, M. J. (1970) Experiments on enzyme induction in *Chlorella pyrenoidosa*. *Journal of Biological Education*, 4, 43–52.

Gordon, A. H. and Eastoe, J. E. (1964) *Practical Chromatographic Techniques*. Newnes, London.

Hageman, R. H. and Flesher, D. (1960) Nitrate reductase activity in corn seedlings as affected by light and nitrate content of nutrient media. *Plant Physiol.*, 35, 700–708.

Hansell, M. H. and Aitken, J. J. (1977) *Experimental Animal Behaviour*. Blackie, London.

Hawcroft. D. M. and Short, K. C. (1973) Some experiments on the light reaction of photosynthesis. *Journal of Biological Education*, 57, 23–26.

Heath, R. W. (1969) A simple experiment to show amylase activity. *School Science Review*, 50 (172), 571–572.

Hewitt, E. J. (1975) Assimilatory nitrate–nitrite reduction. *Ann. Rev. Plant Physiol.*, 26, 73–100.

Jacobson, B. S., Fong, F. and Heath, R. L. (1975) Carbonic anhydrase of spinach. Studies on its location, inhibition and physiological function. *Plant Physiol.*, 55, 468–474.

Kirkby, E. A. and Hughes, A. D. (1970) Some aspects of ammonium and nitrate reduction in plant metabolism. In *Nitrogen Nutrition of the Plant*, ed. Kirkby, E. A., pp. 69–77. University of Leeds.

Klepper, L. and Hagemann, R. H. (1969) The occurrence of nitrate reductase in apple leaves. *Plant Physiol.*, 44, 110–114.

Kramer, P. J. (1955) Water relations of plant cells and tissues. *Ann. Rev. Plant Physiol.*, 6, 253–272.

Mizrahi, Y., Blumenfeld, A. and Richmond, A. E. (1970) Abscisic acid and transpiration in leaves in relation to osmotic root stress. *Plant Physiol.*, 46, 169–171.

Munk, H. (1958) The nitrification of ammonium salts in acid soils. *Landw. Forsch.*, 11, 150–156.

Myers, A. (1981) The blue bottle reaction and photosynthesis. *School Science Review*, 63 (222), 112–117.

Neish, A. C. (1939) Studies on chloroplasts. *Biochem. J.*, 33, 300–308.

Oaks, A., Wallace, W. and Stevens, D. (1972) Synthesis and turnover of nitrate reductase in corn roots. *Plant Physiol.*, 50, 649–654.

Paleg, L. (1960) Physiological effects of gibberellic acid. 1. On the carbohydrate metabolism and amylase activity of the barley endosperm. *Plant Physiol.*, 35, 243–299.

Park, T. (1934) Observations on the general biology of the flour beetle (*Tribolium confusum*). *Rev. Biol.*, 9, 36–54.

Randerath, K. (1966) *Thin-layer Chromatography*. Academic Press, New York and London.

Roberts, J. and Whitehouse, D. G. (1976) *Practical Plant Physiology*. Longmans, London and New York.

Sanderson, G. W. and Cocking, E. C. (1964) Enzymic assimilation of nitrate in tomato plants. 1. Reduction of nitrate to nitrite. *Plant Physiol.*, 39, 416–422.

Slayter, R. O. (1967) *Plant–Water Relationships*. Academic Press, London.

Stevens, I. H. (1981) Polyphenol oxidase – an easily available enzyme. *School Science Review*, 63 (222), 103–105.

Stock, R. and Rice, C. B. F. (1972) *Chromatographic Methods*. Chapman and Hall, London.

Stoneman, C. F. (1972) The action of a proteolytic enzyme. *School Science Review*, **53** (185), 746–747.

Sulebele, G. A. and Rege, D. V. (1967) Temperature sensitivity of catalase induction in *S. cerevisiae*. *Enzymologia*, **33**, 354–360.

Sutcliffe, J. F. (1968) *Plants and Water*. Edward Arnold, London.

Varner, J., Ram Chamdra, G. and Chrispeels, M. J. (1965) Gibberellic acid controlled synthesis of α-amylase in barley endosperm. *J. Cell Comp. Physiol.*, **66** (supp.), 55–68.

Walker, J. R. L. (1975) *The Biology of Plant Phenolics*. Edward Arnold, London.

Wallace, W. and Pate, J. S. (1965) Nitrate assimilation in the field pea (*Pisum arvense L.*) *Ann. Bot. (London) N.S.*, **29**, 655–671.

Wilkinson, J. R. and Wellburn, A. R. (1981) Rapid separation of plant pigments by thin layer chromatography using icing sugar-coated microscope slides. *School Science Review*, **62** (218), 81–83.

Woldendorp, J. W. (1968) Losses of soil nitrogen. *Stikstof Dutch Nitrogenous Fertiliser Review*, **12**, 32–46.

Wyatt, H. V. (1974) Further experiments with digestive enzymes in freshly killed animals. *Journal of Biological Education*, **8**, 330–332.

Names and Addresses of Suppliers

1 B.D.H. Chemicals Ltd
Poole
Dorset
BH12 4NN

2 Diffusion Systems Ltd
45 Rosebank Road
London
W7 2EW

3 Gerrard Biological Centre
Worthing Road
East Preston
West Sussex
BN16 1AS

4 Philip Harris Biological Ltd
Oldmixon
Weston-super-Mare
Avon
BS24 9BJ

5 Arnold R. Horwell Ltd
2, Grangeway
Kilburn High Road
London
NW6 2BP

6 Searle Diagnostic Ltd
Lane End Road
High Wycombe
Buckinghamshire
HP12 4HL

7 Sigma (London) Chemicals Co. Ltd
Fancy Road
Poole
Dorset
BH17 7NH

8 Rank Strand Ltd
P.O. Box 24
Mitchelston Industrial Estate
Kirkcaldy
Fife
Scotland
KY1 3LY

Index